A Diversity of Pathways Through Science Education

Yann Shiou Ong · Timothy Ter Ming Tan ·
Yew-Jin Lee
Editors

A Diversity of Pathways Through Science Education

Springer

Editors
Yann Shiou Ong
National Institute of Education
Natural Sciences and Science Education
Nanyang Technological University
Singapore, Singapore

Timothy Ter Ming Tan
National Institute of Education
Natural Sciences and Science Education
Nanyang Technological University
Singapore, Singapore

Yew-Jin Lee
National Institute of Education
Natural Sciences and Science Education
Nanyang Technological University
Singapore, Singapore

ISBN 978-981-97-2606-6 ISBN 978-981-97-2607-3 (eBook)
https://doi.org/10.1007/978-981-97-2607-3

This Springer imprint is published by the registered company Springer Nature Singapore Pte Ltd.
The registered company address is: 152 Beach Road, #21-01/04 Gateway East, Singapore 189721, Singapore

Paper in this product is recyclable.

Introduction

ISEC 2020 was intended to be *the* conference for showcasing research and new thinking in science education. Organized once every couple of years by staff from the Natural Sciences and Science Education academic group at the National Institute of Education, Singapore, its opening was initially planned for the middle of 2020 to coincide with the mid-year or summer holidays for local and international audiences. The International Science Education Conference (ISEC) organizing committee expressed high hopes for at least two reasons: (i) STEM education was to assume an equally prominent theme during this conference, and (ii) ISEC-STEM 2020 aspired that its attendees experience fresh insights on current/future trends and needs in these domains arising from 20-20 vision in this auspicious year. As such, the overall conference theme was entitled *The Tango between Science and STEM* to reflect these aforementioned ideas. It was described on the official website as being able to

> reflect a dance the S-T-E-M education researchers are immersed in as they crossover into interdisciplinary research. Tango also encapsulates the synergy between Science and STEM as Science continues to play a prominent role in STEM education. One of the disruptions to science education as a field is the increasing emphasis on integrated STEM education. With science as the discipline that is currently dominant in integrated STEM, it is strategic that we position ISEC 2021 and STEM 2020 as two related conferences. This will encourage scholars from both fields to interact and develop synergies to move the knowledge forward. (from the conference website)

A disruption of immense proportions did indeed occur although not in the way that the organizers had anticipated because the COVID-19 pandemic plunged most of the world including Singapore into disarray soon after the winter of 2019. This global event scuttled our conference planning resulting in ISEC-STEM being delayed for a year to the summer of 2021 as well as being conducted virtually. STEM 2020 was also decoupled from ISEC 2021 due to various considerations by the organizers. These changes were not as bad as was thought, for now the problems of high registration fees as well as long-distance travel woes were overcome at one stroke though not the issue of participating in real-time across very different time zones. The ISEC 2021 conference was nonetheless successful under these very difficult circumstances with 77 papers/symposia presented by researchers and teachers from 17 states/regions. As with previous ISEC, presenters from Singapore occupied the lion's share of presentations. What readers therefore see in this edited book are a sampling of invited authors who had presented at ISEC 2021.

This book is organized into three parts.

1. Part I: Questions and Questioning in Science/STEM Education
2. Part II: Developing Science Teaching and Assessment
3. Part III: History, Philosophy and Sociology of Science/Engineering and Informal Learning

To summarize, Part I features three chapters foregrounding the epistemic practice of student questioning across grade levels. Part II is hugely diverse in its coverage with five chapters describing different aspects of teaching, learning, and assessment from multiple theoretical standpoints while Part III comprises three chapters that also appear to be very diverse but can be seen as takes on the history and/or development of formal and informal learning in science and engineering. The beginning of each part is accompanied by a Commentary written by each of the book editors who were members of the ISEC organizing committee.

We wish to bring to the reader's attention a unique feature of this edited book: At the end of a chapter, each set of authors has written a "Note to Future Colleagues" to describe their aspirations for the state of science/STEM education research in 2050 in the research area reported in their chapter. Collectively, these "Notes" point toward potential directions that science/STEM education research could take to achieve the espoused visions by the middle of the twenty-first century. On this note (pun intended), the editors of this book would like to end this introduction with our own Note to Our Future Colleagues in 2050.

Note to Our Future Colleagues

It is the year 2050. According to authors Christiana Figueres and Tom Rivett-Carnac of the book *The Future We Choose*, if the world worked together and took appropriate actions to avert the climate crisis, we would be on track to warm by no more than 1.5 °C by the year 2100 (Figueres and Rivett-Carnac 2020). Thus, our future looks bright and science/STEM education continues to be an important part of our culture. Here, we present our three predictions made in the 2020s for science/STEM education in 2050.

I. Epistemic Practices of the Future

By 2050, disciplinary boundaries have blurred as most professionals work in interdisciplinary teams to solve the complex real-world problems. Using the language of 2020s, most professionals and experts are trained in one or more of the "traditional disciplines" such as the sciences or humanities and have working knowledge of several other disciplines. Thus, educators and education researchers have progressed from focusing on disciplinary to interdisciplinary epistemic practices. As working in large interdisciplinary teams is the norm of work in 2050, interpersonal or "soft" skills such as collaboration, communication and empathy have become just as important as content knowledge and procedural knowledge. In addition, engagement in epistemic practices has become part of school norm (though schools no longer take the same form as they did in the 2020s). The ability to critique, construct, and discern trustworthy knowledge claims is now essential to everyday living since most people have become content creators as well as content consumers. Since everyone can find a platform to publish their views, some of which are erroneously/intentionally positioned as truths, it has become challenging—yet part of everyday life—for the layperson to discern trustworthy and sound claims from unwarranted ones, including scams. Incidentally, attempts by large-scale social media platforms and governments to curate online information have failed and thus, the onus of fact-checking remains on the individuals.

II. Science Teaching, Learning, and Assessment in the Age of AI

In the coming three decades, science teaching and learning will become both easier and harder. It will seem easier because so much more will be known about the overall principles of how human cognition functions in the service of acquiring valuable knowledge. The main theories or frameworks will have been mapped out regarding how cognition is dependent on one's internal architecture of neurons as well as the body's engagement with its contextual surround. On the other hand, how cognition interacts with other aspects such as the physical body and its observable states known as emotions or affect will have complexified the fine details of how people learn and behave. Besides, schooling will now be augmented by forms of artificial minds or intelligences in a similar manner as how tools of the past (e.g., the abacus, counting rods, log tables, calculators, the Internet) had assisted classroom learning. The nature and target of assessment too will likely change drastically as mentioned in the commentary for Part II. So despite knowing more than ever about how people learn and in possession of unimaginable new technologies, science teachers in the middle of the century will still have their work cut out for them as they facilitate students in much more demanding tasks (i.e., see the above epistemic decision making) compared to previous eras. A teacher's life will probably remain just as demanding rather than easier and the space-age life with robot teachers imagined in the US cartoon *The Jetsons* will not materialize.

III. Schools of the Future

The COVID-19 pandemic was the trigger that kickstarted what will likely be the biggest evolution in schools since the widespread adoption of formal schooling during the first Industrial Revolution, and that 30 years hence, its repercussions continue to shape and redefine what "schools" are. The lockdowns and restrictions imposed certainly advanced the pervasive use of and drove the development of technologies for remote learning and remote work. More significantly, these affordances for and the experience of remote learning normalized it as another mode of "school". The evolution began with home-based learning being instituted on a routine basis, where students stay home one day a month or fortnight, mainly as a way to familiarize students and teachers with remote learning as a contingency against subsequent pandemic-induced disruptions as has already happened in Singapore during the pandemic (Tan and Chua 2021). We see schools increasingly becoming delocalized in subsequent decades, where students can connect to their classes remotely or attend in-person as circumstances or preference dictate. Since the home environment may not always be the most conducive for learning, co-learning spaces will become commonplace. Modeled after coworking spaces where companies lease office space or traveling workers rent a desk for a day or two, students settle into an individual "pod" or as a small group of peers in mini classrooms to attend their lessons on the other side of town, the country, or the world. Students from small remote communities or those from impoverished neighborhoods can receive a quality education at bigger schools without geographical constraints. Inter-school collaborative learning, up to and including those at an international level, is not uncommon and allows for cross-cultural learning that promotes pluralistic understanding and empathy. The co-learning spaces may blend formal and informal learning opportunities, synchronous and asynchronous learning modalities, as well as provide the socialization and interaction among peers that home-based learning does not. As mentioned above, the demands on teachers in 2050 won't be easy. Technologies would certainly have been developed to facilitate and enable more naturalistic remote presence and interactivity between teachers and learners. Of particular relevance to us would be the ways in which science practical work might be conducted in such settings. Perhaps such co-learning spaces would be co-located with community libraries and science centers/museums to jointly form satellite venues for both formal and informal learning. But most importantly for such *delocalized schooling* to have happened successfully, researchers and practitioners must have studied, developed, and refined the pedagogies and management techniques for hybrid classes where some students are physically present while some are remotely connected.

While all these prospects may be futuristic to us in the 2020s, we are confident that you, our colleagues in educational research, have continued to study teaching and learning as a learner-centered and hence human endeavor.

References

Figueres C, Rivett-Carnac T (2020) The future we choose: surviving the climate crisis (1st edn). Alfred A. Knopf

Tan OS, Chua JJE (2021) Science, social responsibility, and education: the experience of Singapore during the COVID-19 pandemic. In: Reimers FM (ed) Primary and secondary education during Covid-19. Springer

Contents

Part I
Questions and Questioning in Science/ STEM Education

Chapter 1
Commentary for Part I: Educating Students for Good Questioning in Science/STEM

Yann Shiou Ong

A common thread across the three chapters in Part I is the practice of *good questioning* in science/STEM. Chapter 2 by Uchinokura, Kusuhata and Hiroshi and Chap. 3 by Regunathan, Tan, and Koh discuss the practice of students asking/posing questions. Chapter 4 by Leung involves deliberating the trustworthiness of reports on science-related issues. I assert that the deliberation of trustworthiness requires good questioning as it involves asking critical questions (Walton & Godden 2005; Walton 1996) to elicit information from such reports that help determine their trustworthiness and subsequently, the reader's placement of trust. Examples of such critical questions include: who are the authors/funders of the research and do they have a biased agenda? Is the research study reliable and valid? Is the claim/conclusion sound, based on the research design that includes methods of investigation/data collection and data analysis? What evidence do the authors have and are they sufficient to support the claim/conclusion? Critical questions could be asked inter-personally of one another, such as during peer discussion, or intra-personally as questions on our mind as we review the reports.

1.1 Questioning as an Epistemic Practice

Questioning can be considered an epistemic practice, which is "a socially established set of activities directed towards common epistemic goals" (Watson 2018, p. 78). To elaborate, drawing upon Kelly and Licona's (2018) interpretation of epistemic practices, framing good questioning as an *epistemic practice*–rather than a process skill or method in science/STEM–implies that good questioning is interactional (observed

Y. S. Ong (✉)
National Institute of Education, Nanyang Technological University, Singapore, Singapore
e-mail: yannshiou.ong@nie.edu.sg

Y. S. Ong et al. (eds.), *A Diversity of Pathways through Science Education*,
https://doi.org/10.1007/978-981-97-2607-3_1

among people of a community through their coordinated activities), contextual (situated within sociocultural norms and practices), intertextual (involves communication of discourses including signs and symbols in a historically coherent way e.g., past discourses have bearing on present questioning), and consequential (legitimized knowledge established through questioning exemplifies power and culture of the community). The epistemic goal of questioning is to seek out information to help us form trustworthy beliefs and decide how to act (Watson 2018). As a practice cannot be sufficiently defined by a finite set of rules and descriptions to effectively guide the reasoning and action necessary to enact the practice (Ford 2015), framing questioning as a practice rather than a process skill thus implies that, just like other practices, it is operationally challenging and perhaps near impossible to describe in reasonable details how to ask questions.

1.2 Good Questioning Versus Asking Questions

Readers familiar with the practice of "asking questions" might wonder if "good questioning" is just a synonym of the former. Afterall, "asking questions" as elaborated in the USA Next Generation Science Standards (NGSS) (NGSS Lead States 2013) and "posing questions" in the Singapore Ministry of Education's interpretation of scientific practices i.e., Ways of Thinking and Doing in Science (MOE 2021) seem to describe the same practice, for example. However, the difference between asking question and good questioning is non-trivial. Asking questions implies the competency to elicit any information. On the other hand, good questioning requires the competency to elicit *worthwhile* information through making appropriate judgements about what information to elicit, from whom, when, and how to do so (Watson 2018). Of course, not all questions are asked with eliciting information as the goal. For example, during argumentation, questions can be asked to elicit information from someone with the goal of challenging their claim or justification and thus casting doubt on their argument. Nevertheless, such questions still function as a speech act to elicit information from someone.

1.3 Why Should We Educate Students for Good Questioning in Science/STEM

Since "[q]uestioning initiates, guides and shapes inquiry" (Watson 2018, p. 8), it stands to reason that questioning likewise initiates, guides, and shapes disciplinary inquiry such as scientific inquiry as well as interdisciplinary inquiry such as STEM problem-solving. Good questioning perhaps also accounts for how serendipity plays a part in some scientific discoveries. As Louis Pasteur aptly put it, "in the fields of observation, chance only favours the mind that is prepared" (Vallery-Radot & Devonshire, 1917). When the opportunity presents itself, it is only the prepared mind of a good questioner with deep, disciplinary expert knowledge who can competently elicit worthwhile information from nature i.e. get nature to "speak" (Ford 2008. p. 408) that will end up pursuing an intellectually virtuous or meaningful line of inquiry. Even if such a line of inquiry does not lead to success, there will be much to be learnt from the failure. The above rings true even for more mundane, "business-as-usual" lines of inquiry. Thus, good questioning plays a significant role in the construction of knowledge claims/solutions to problems. Additionally, good questioning during argumentation in challenging someone's proposed argument, especially in light of epistemic criteria or ideals and reliable processes for arriving at the (inter)disciplinary knowledge claims, contributes towards the peer review process which is the current gatekeeping mechanism for knowledge claims. Thus, good questioning arguably supports both the goals of construction and critique of knowledge claims (Ford 2008), which in turn correspond to the intellectual virtues of inquisitiveness and healthy scepticism.

1.4 What the Chapters Tell Us About Questions and Questioning

Collectively, the three chapters in this section spanned the practice of questioning across the elementary, secondary, and post-secondary education levels. Chapters 2 and 3 focused on students' epistemic practice of asking questions at different grade levels and in different contexts. Both studies focused on the nature or type of questions asked by students, that is , what counts as good questions. Chapter 2 by Uchinokura, Kusuhata and Hiroshi determined fourth versus sixth grade elementary students' understanding of researchable questions in science, i.e., questions appropriate for answering through scientific inquiry, by asking students to identify researchable from non-researchable biology related questions and to generate their own questions based on a physics experiment. The researchers found that while it was fairly easy for the students to identify the researchable questions, only a limited number of students could pose researchable questions i.e., *what* and *how* questions; most students asked *why* and closed questions unsuitable for answering through science inquiry. In Chap. 3, Regunathan, Tan, and Koh compared and contrasted middle school students' questioning patterns during group presentation of their solution to a STEM problem after experiencing either the problem-centric approach or solution-centric approach integrated STEM instruction. The researchers did not find any statistically significant differences in the questions raised by students who experienced the different integrated STEM instruction. However, they noted that ontic questions (questions foregrounding the nature of entities, such as "What is a force?" or "What is the difference between mass and weight?") were asked most frequently, followed by causal questions, while epistemic questions seeking justifications and understanding of how we know what we know and why we should trust a claim, were posed the least. Finally, Chap. 4 by Leung described how a lesson unit designed using a practice approach to the nature of science (NOS) fostered Hong Kong SAR undergraduate students' scepticism towards placement of trust in science experts towards science-related social issues. These students were enrolled in a general education course and thus, not all of them have strong science background. Leung found an increased in the students' understanding and concern about trust placement as well as their understanding of NOS as indicated in students' reflective journals. However, some challenges remained in the students' trust placement.

1.5 Questioning in the Future of Science/STEM Education

In their **Notes to Our Future Colleagues**, the authors of the three chapters have depicted a future where integrated STEM teaching and learning is the default classroom practice, with students and teachers working toward creating solutions for real world problems (Regunathan and colleagues, Chap. 3). As such, students' questions-driven curricula becomes a reality, as students pose the questions they would like to explore and answer—many of which would be interdisciplinary in nature—through open inquiry approaches as students work collaboratively within learning communities comprising peers and students of different age groups, parents, teachers, and other experts to learn the necessary conceptual, procedural, epistemic, and social knowledges in order to answer their questions (Uchinokura and colleagues, Chap. 2). For students to be competent in directing their learning against the backdrop of the infodemic in the future, they need to achieve a good grasp of the epistemic practices in various disciplines in order to successfully distinguish trustworthy information from misinformation and disinformation. The placement of trust does not merely involve rationale thinking. Emotions also play an important role in the placement of epistemic trust on knowledge claims as emotions could mediate the cognitive, metacognitive, motivational and social aspects of trust placement (Leung, Chap. 4). This leads to the question of whether good questioning–including asking of critical questions to elicit worthwhile information–might also be mitigated by emotions and to what extent.

In our pursuit to develop students' ability to ask good and meaningful questions in science/STEM learning activities, educators need to be mindful not to only develop merely the skill of asking questions, but rather, the epistemic practice of asking good questions in science/STEM. Practice denotes a way of being, a disposition, or habit of mind. We ask questions because it is the appropriate act in a given context, such as when we have a need to know more, or when we have doubts about the soundness or trustworthiness of a claim/argument. There also needs to be an understanding of what "good" questions in science/STEM look like. For example, as articulated in the US NGSS, good scientific questions in the K-12 setting are questions that are researchable or can yield answers through science inquiry in the classroom. Such questions were the target of research by Uchinokura and colleagues (Chap. 2). And what about good questions in STEM activities? While Regunathan and colleagues (Chap. 3) categorised the nature of students' questions related to STEM solutions as ontic, casual or epistemic, are all three types of questions equally good? Is it desirable to have students pose a mix of all these types of questions or should some types of questions be prioritised over others? And to achieve what goals?

In closing this commentary, I propose two question prompts for readers to consider, with the intention of generating some worthwhile research questions for future inquiry into good questioning. Firstly, should instruction aimed at developing students' practice of good questioning for construction/critique of knowledge claims be well-structured (e.g., using the Question Formulation Technique created by the Right Question Institute (rightquestion.org)) or open-ended (e.g., students ask questions freely)? Or perhaps a progression from structured to open-ended as scaffolding is provided then faded? Secondly, on questions for critique of knowledge claims, how should students be taught to frame critique questions/critical questions to minimize the loss of face when critiquing peers' ideas (or even the teacher's ideas!) in cultures where face and respect for authority are highly valued? I invite readers to take on the challenge of answering either of the two questions I have posed, or perhaps be inspired to generate further good questions for science/STEM education research communities to pursue after reading this commentary.

References

Ford M (2008) Disciplinary authority and accountability in scientific practice and learning. Sci Educ 92(3):404–423. https://doi.org/10.1002/sce.20263

Ford MJ (2015) Educational implications of choosing "practice" to describe science in the next generation science standards. Sci Educ 99(6):1041–1048. https://doi.org/10.1002/sce.21188

Kelly GJ, Licona P (2018) Epistemic practices and science education. In: Science: philosophy, history and education. Springer Nature, pp 139–165. https://doi.org/10.1007/978-3-319-62616-1_5

Ministry of Education [MOE] (2021) Science teaching and learning syllabus—Lower secondary express course and normal (academic) course. Ministry of Education, Singapore

NGSS Lead States (2013) Next generation science standards: for states, by states. The National Academies Press

Walton D (1996) Argument schemes for presumptive reasoning. Erlbaum, Mahwah, NJ

Walton D, Godden D (2005) The nature and status of critical questions in argumentation schemes. In: Hitchcock D (Ed) The uses of argument: proceedings of a conference at McMaster University. Ontario Society for the Study of Argumentation, pp 476–484

Watson L (2018) Educating for good questioning: a tool for intellectual virtues education. Acta Analytica 33(3):353–370. https://doi.org/10.1007/s12136-018-0350-y

Chapter 2
Primary School Students' Understanding of Posing Questions for Scientific Inquiry

Shingo Uchinokura, Misato Kusuhata, and Naoya Hiroshi

2.1 Introduction

A cognitive tool that is frequently used in education and is important for teaching and learning science for all grades in schools is asking questions. Both teachers and students have opportunities to ask questions in the classroom. Based on observational studies conducted in high school classrooms, teachers often ask students more questions during lessons, as compared to students (Dillon 1988). While teaching science, teachers ask questions to determine the students' understanding in formative assessments, evaluate higher-order thinking skills, stimulate inquiry into the natural world through investigations, encourage problem-based learning and practical work, and invoke their reflection on practices followed during lessons (Chin 2006, 2007; Chin and Osborne 2008; Oliveira 2010). From the perspective of social language, classroom discourse is controlled by teachers' speech acts, which include asking questions (Mehan 1979; Lemke 1990).

However, although the number of students asking questions depends on their age, and the fact that older students seldom ask questions (Dillon 1988; Osborne and Reigh 2020), students asking questions should not be so inadequate as to impede their learning of science. Students' questions contribute to constructing knowledge, pursuing investigations, fostering discussion and debates in classroom discourses, monitoring and self-evaluating their understanding, and increasing motivation and interest to understand a topic (Blonder et al. 2008; Chin and Brown 2002; Chin and

S. Uchinokura (✉) · M. Kusuhata
Faculty of Education, Kagoshima University, Kagoshima, Japan
e-mail: uchinokura@edu.kagoshima-u.ac.jp

M. Kusuhata
e-mail: k3970101@kadai.jp

N. Hiroshi
Attached Junior High-School, Kagoshima University, Kagoshima, Japan
e-mail: rgf171@edu.kagoshima-u.ac.jp

Y. S. Ong et al. (eds.), *A Diversity of Pathways through Science Education*,
https://doi.org/10.1007/978-981-97-2607-3_2

Osborne 2008; Herranen and Aksela 2019; Kuhn et al. 2020; Osborne and Reigh 2020). The types of questions that students ask while learning science were identified through classroom observations during lessons and interviews, before and after instruction (Chin and Brown 2002). These were categorised into basic information questions and wonderment questions. The basic information questions included factual or procedural questions. Factual questions usually only require the recall of information and are often closed questions, questions to clarify a given procedure, or questions on how a task is to be carried out. Wonderment questions require an application or extension of taught ideas, and focus on predictions, explanations, and causes, or resolving discrepancies and gaps in knowledge. The type of written questions students asks, after reading chapters from their textbook or scientific articles, were also explored (Hofstein et al. 2005; Marbach-Ad and Sokolove 2000). The questions ranged from low-level questions, related to facts and explanations of the phenomena, to high-level questions that could be answered only after further investigation.

2.1.1 Students' Questions for Scientific Inquiry

Amongst different types of questions that students ask while learning science, as abovementioned, inquiry-based or researchable questions have been specifically emphasised in inquiry-based science education (Herranen and Aksela 2019; Rönnebeck et al. 2016). The sciences are perceived as a practice that seeks to answer three overarching questions: *ontological*, *causal*, and *epistemic* (Osborne and Reigh 2020). An ontological question is a simple ontic question about observable entities, like 'What exists?'. A causal question seeks casual and mechanical explanations for the natural world. These are questions like 'What causes A?'. An epistemic question seeks reasons to justify scientific knowledge and argumentation, as well as believe them, like 'How do we know?' and 'Why should we believe this?'. These questions reflect scientific inquiry as an epistemic practice. The significance of asking questions in science learning is elaborated as follows (Herranen and Aksela 2019): Posing questions is an inquiry skill, and the questions are part of the inquiry. Inquiry-based science, which is considered a more authentic learning experience compared to superficial learning such as memorizing factual knowledge, is usually initiated by posing questions. Asking questions enhances students' motivation, interest, responsibility, or ownership of learning, as these are driving forces for inquiry. This is closely associated with the development of thinking, articulation, argumentation, and reasoning skills. In addition, the ability to ask questions, that is, problem finding or problem formulation in other contexts, could also be seen as a predictor of creative accomplishments (Runco 1994; Abdulla et al. 2018). In response to these discussions or competency-based science education reforms, developing the ability to ask questions and engaging students in scientific inquiry are emphasised in the science curricula in various countries, including USA and Japan. (Achieve 2013; National Research Council 2012, 2013; Ministry of Education, Culture, Sports, Science and Technology 2017a, b). These science curricula expect students not only to ask the lower-level type of information-seeking questions, but also to identify and ask researchable questions for practice during science learning. Indeed, asking questions is perceived as an important scientific practice.

Formulated questions for scientific inquiry are defined as specific types of questions that can be answered through scientific practices, including observation and experimentation (Cuccio-Schirripa and Steiner 2000; Hartford and Good 1982; Keys 1998; Allison and Shrigley 1986). Alfke (1974) proposed the idea of operational questions that support teachers and students in investigating scientific phenomena. The use of operational questions includes the manipulation of variables by eliminating, substituting, and/or increasing or decreasing the presence of a variable. By measuring the selected variables, through close observation and interpretation of the evidence, students can answer these questions (Allison and Shrigley 1986). Questions asked about the natural world can be classified as researchable or non-researchable. Some non-researchable questions include questions that students cannot collect data about because of a lack of content and procedural knowledge. In addition to enhancing the proficiency level of the science curriculum, resource issues in the science learning environment, such as better experimental equipment and facilities, are needed. However, there are certain non-researchable or improper questions for scientific inquiry that cannot be addressed even by experts. For example, questions related to personal preferences and ethical or political judgments are not suitable for natural science.

Studies have focused on the ability of high school students to ask better researchable questions and to refine their questions for further research (Blonder et al. 2005; Dori and Herscovitz 1999; Hofstein et al. 2004; Lombard and Schneider 2013). The participants in these studies were students participating in scientific research projects or advanced courses in which they could learn to pose questions in more authentic ways. The results indicate that open-inquiry activities have a positive effect on the way questions are asked. The questions asked by students through inquiry-based learning improved both quantitatively and qualitatively (Hofstein et al. 2004). In terms of the correlations between students' achievements and the level of their inquiry questions, while average and below-average performance students attempted to understand the subject matter to ask inquiry questions and hypothesise accordingly, high-achieving students attempted to ask questions on new things to apply the knowledge to other situations (Blonder et al. 2008). Research has been conducted on the quantity and quality of questions in practical work, such as during guided inquiry and problem-solving (Keys 1998; Otawa and Kinoshita 2014; Sakamoto et al. 2016). Keys (1998) reported that many students from a middle school in the United States formulated researchable questions that explored the effect of a variable on a scientific event through the generative model for teaching primary sciences, including those associated with the exploration, investigation, and reflection phases. Studies have also examined the impact of direct instruction on posing questions in primary school (Chin and Kayalvizhi 2002; Allison and Shrigley 1986), middle school (Cuccio-Schirripa and Steiner 2000), and high school (Hartford and Good 1982). Allison and Shrigley (1986) reported that students who learned through the teacher modelling of asking operational questions asked more operational questions than students who did not have this experience. While students in the group collaboratively asked researchable questions after some examples of questions had been shared with them, they faced difficulties asking proper questions individually without instructional modelling (Chin and Kayalvizhi 2002). It has been reported that the ability to ask questions is malleable; if students are explicitly taught how to ask and answer questions, they will progress better (Osborne and Reigh 2020).

There are different types of questions for inquiry; accordingly, taxonomies of researchable questions were proposed (Chin and Kayalvizhi, 2002; Krajcik et al. 1999). The questions asked by students from different schools range from researchable questions to non-researchable ones. Chin and Kayalvizhi (2002) reported that when a teacher asked sixth-grade students from a primary school to write down questions, they preferred to investigate in different scientific subfields individually; very few of their questions were identified as researchable. Instructional scaffoldings included showing examples of questions and talking with other students, which resulted in them being able to omit the non-researchable questions, and then collecting more researchable questions than in their first attempt. By categorising the students' questions, Chin and Kayalvizhi (2002) proposed a taxonomy of researchable questions: *comparison, cause-and-effect, prediction, design-and-make, exploratory, descriptive, pattern-seeking* etc. Their categories of students' questions showed the posed questions that guide the scientific inquiry and the expected types of answers to each question. On the other hand, Krajcik et al. (1999) proposed a simpler and more comprehensive categorization of posing questions for scientific inquiry: *descriptive*, *relational*, and *cause-and-effect questions.* In their definition, *descriptive questions* allow students to find out observable characteristics of phenomena, *relational questions* allow students to find out about associations between the characteristics of different phenomena, and *cause-and-effect questions* allow students to make inferences about how one variable affects another variable (Krajcik et al. 1999). As examples, "What materials dissolve in water?" is a *descriptive question*, "Which dissolves faster in water, salt of sugar? is a *relational question*, and "How does the temperature of water affect materials dissolves in water?" is a *cause-and-effect question.* Ontological questions in the more recent types of questions proposed by Osborne and Reigh (2020) have correspondence to descriptive and relational questions, and casual questions is the same as cause-and-effect questions by Krajcik et al. (1999). Epistemic questions are the meta-science questions of validation for these researchable questions. In this chapter, the framework of researchable questions proposed by Krajcik et al. (1999) is followed because the taxonomy of researchable questions by Chin and Kayalvizhi (2002) corresponded to the framework by Krajcik et al. (1999) and the framework could be expected to explain the types of researchable questions economically.

Hiroshi and Uchinokura (2019) reported that the skills of Japanese junior high school students associated with posing and identifying researchable questions could be differentiated according to the types of questions asked. In their study, the students were asked to answer whether each statement that mentioned the natural and mechanical world could be confirmed through investigation. If the statement was scientifically confirmable, it had to be changed from a declarative sentence to a question sentence. Identifying and asking *relational questions* was cognitively more difficult for them than *descriptive questions*. For the former type, they asked *closed questions* with binary answers without sufficient awareness of the criteria for comparison and reference. Most of the students asked questions using interrogatives that are specific words that have a grammatical function of focusing on subjects, objects, place, selection, amount, reasons and so on, such as 'why' and 'what'. These asked questions were interpreted as questions about the ontological meanings and backgrounds of natural things and events. However, students could differentiate researchable questions from non-researchable ones. This implies that they have epistemic knowledge of the two. There is limited research on this topic among primary school students. Additionally, it has not been determined whether there are any differences in the quantity and quality of questions asked by students of different grades. Gender differences in identifying scientific issues among older students have been reported (Organisation for Economic Cooperation and Development 2015). This study poses the following three research questions: How well do primary school students identify researchable questions? What kinds of questions (i.e. researchable or non-researchable) do students ask based on a given situation? Are there any differences in the understanding of posing questions for inquiry between fourth and sixth-graders and between male students and female students?

2.2 Methods

2.2.1 Participants

The participants of the study reported in this chapter comprised 99 fourth-grade students (8–9 years old), including 53 males and 46 females from three classes, and 98 sixth-grade students (11–12 years old), including 55 males and 43 females from three classes. The participants were from the same public elementary school in Japan and were taught by the same teacher. All fourth-grade and sixth-grade students who attended the school participated in the study, except the absentees. The national science curriculum in Japan, that is, the *Course of Study (CoS)*, was developed by the Ministry of Education, Culture, Sports, Science, and Technology (MEXT) and is revised approximately every ten years. Students in both grades were taught according to the 2008 revised *CoS* (MEXT 2008), which focuses on nurturing the ability to think, decide, and express. This was emphasised at all school levels from kindergarten to upper secondary school (Matsubara 2018), and according to the *CoS*, the sciences include physics, chemistry, biology, and earth and space science. The curriculum for fourth-grade students includes various topics such as simple electric circuits, changes in the state of matter, the human body, seasons, and the moon and stars.

For sixth-grade students, topics include mechanics, electricity, combustion, solution, photosynthesis, and weather. However, the nature of scientific inquiry as a topic of epistemic knowledge is not taught explicitly in the national science curriculum, including the role of posing questions in scientific inquiry and the differentiation of researchable and non-researchable questions.

2.2.2 Data Collection and Analysis

To determine the understanding of primary school students regarding posing questions for inquiry in a topic-related manner, a paper–pencil test was developed. This included two types of assignments, which had to be completed without the teacher's help, other than clarifying any questions that students had about the assignments. In the first assignment, there were two items whereby the students were asked to identify which is the better question for further inquiry from among the two given questions and to provide reasons for their choice. In the second assignment, there was one item whereby the students were asked to pose as many questions for scientific inquiry in the given context as they could. In the first assignment, to determine their level of understanding, two items in which students had to differentiate researchable from non-researchable questions in the subject of biology were prepared. The first item to identify researchable questions asked about butterflies visiting the cabbage field (Item 1. *Butterflies in cabbage field*). This was selected to determine their understanding of *descriptive questions* and those using interrogative words like 'why'. The non-researchable question was, 'Why did butterflies come to the cabbage field today?' This could be interpreted as inquiring about the ontological meaning and background of natural things and events. The second item inquired was about the colour differences in fish in the shallow and deep-sea areas (Item 2. *Fish in shallow and deep-sea areas*; Fig. 2.1). This was selected to judge their understanding of *relational questions* and how students apply the epistemic knowledge of posing questions for inquiry to an unfamiliar situation in the primary school science curriculum. In the item, the non-researchable question was, 'Why do fish that live in the deep sea look worse than fish that live in the shallow sea?' This could be interpreted as inquiring about the appearance of fish based on personal preferences. The students were asked to choose a better question for further inquiry and list their reasons.

In the second assignment, to examine students' ability to ask researchable questions based on preliminary observations, a topic based on physics was identified. Students were asked to observe the temperature change of a solution by heating it in a microwave as depicted in Fig. 2.2. Note that the heating experiment was uncontrolled. Through an intentionally created unstructured situation, students had the opportunity to investigate different variables related to changes in temperature. They were encouraged to ask as many questions as possible. This assisted in determining the number and types of questions that students ask by focusing on the usage of interrogative words. The expected researchable questions in this context were *relational questions* that students find out associations between the temperature of the solution

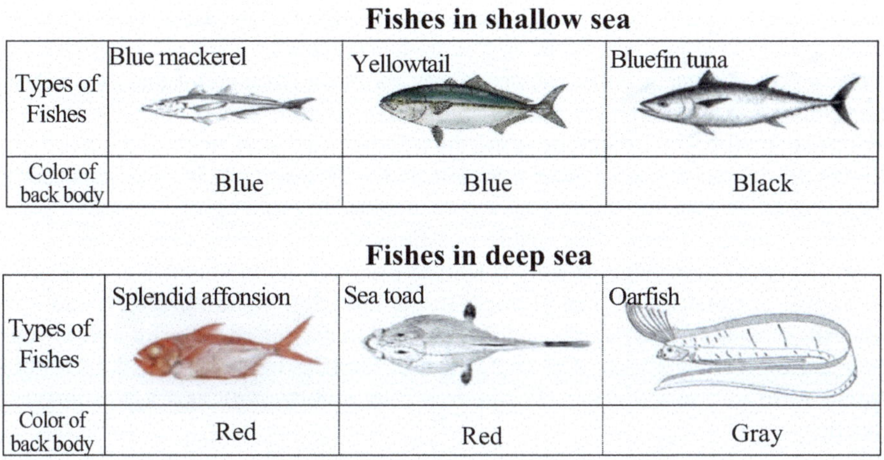

Fig. 2.1 Fish in shallow and deep sea around Japan

and the solute, amount, position, or container of the solution, and *cause-and-effect questions* that students make inferences about how one variable of the solution affects to the temperature of the solution.

All the participants completed the paper–pencil test. Structured interviews, whose duration ranged from 10 to 15 min, were conducted with 12 students from each grade after the test. The interviewed students were five males and seven females from the

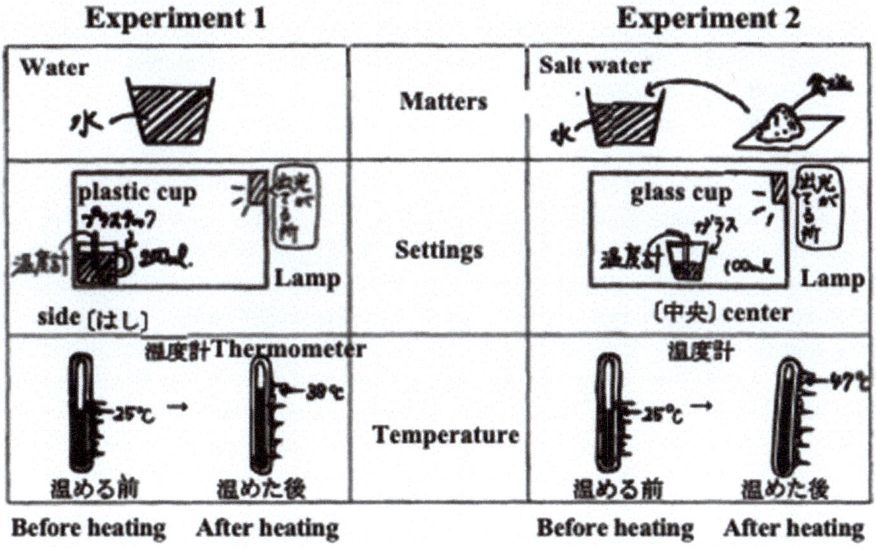

Fig. 2.2 Heating a solution in the microwave

fourth grade, and six males and six females from the sixth grade. These students were selected by following steps. In the first step, the students were categorized into two different performance groups: one group was above the average for this assignment and another was not. In the second step, more than 10 students were listed from each performance group in both grades randomly. In the final step, approximately equal number of students of each gender were selected by following their science teacher's suggestions. The structured interviews were intended to further probe students' ideas of researchable and non-researchable questions and the nature of science that is a meta-science knowledge of how scientific practices is done and how scientific knowledge is constructed. The protocol in the interview was transcribed into text data. All responses to the paper–pencil test and the transcription of the interview from the interview were analysed. Age and gender differences in posing questions for the inquiry were scrutinised using *chi-square testing* and *t-testing* as a nonparametric statistic.

2.3 Results and Discussion

2.3.1 Students' Ability to Identify Researchable Questions

The students' ability to identify researchable questions was explored based on their responses in the paper–pencil test. The answers in the tests are presented in Table 2.1.

In Item 1, 82% of the fourth-grade students and 82% of the sixth-grade students identified that the question about butterfly behaviour was a researchable question as it did not focus on the intent or timing. The question was considered a descriptive question to identify the observable characteristics of the phenomenon and would serve as a better question for further inquiry than the other option. In terms of gender differences, 83% of the male fourth-grade students and 80% of the female fourth-grade students, and 83% of the male sixth-grade students and 80% of the female sixth-grade students, answered correctly. In Item 2, 77% of the fourth-grade students and 87% of the sixth-grade students stated that the relationship between the depth of the sea and the physical characteristics of the fish that lived there was researchable, rather than the question about their appearance. The former is researchable as it is considered a relational question to make inferences about how one variable affects another. They stated that this would be a better question for further inquiry. In terms of gender differences, 81% of the male fourth-grade students and 72% of the female fourth-grade students, and 87% of the male sixth-grade students and 86% of the female sixth-grade students, answered correctly. Through *chi-square testing*, no statistically significant difference was found in the students' answering patterns for both items, considering the grade level and gender differences.

Here we found that the students could identify the descriptive questions by comparing them with intentional ones and the relational questions by comparing them with personal preferences. Of the total participants, 62% of the fourth graders

Table 2.1 Students' identifying questions as researchable

Item	Fourth grade students (N = 99) Person (%)	Sixth grade students (N = 98) Person (%)	p-value
Item 1. *Butterflies in the cabbage field* Seven butterflies were found in the cabbage field at 10 am today			
A: What were the butterflies doing in the cabbage field? (*descriptive questions*)	81 (82%)	80 (82%)	0.97 *n.s*
B: Why did butterflies come to the cabbage field today? (*non-researchable questions*)	17 (17%)	17 (17%)	
No response	1 (1%)	1 (1%)	
Item 2. *Fish in the shallow and deep sea* Different fish were found in the shallow and deep sea near Kagoshima (Japan) as shown in Table 2.1			
A: Why do fish that live in the deep sea look worse than fish that live in the shallow sea? (*non-researchable questions*)	22 (22%)	12 (12%)	0.70 *n.s*
B: How does the body colour of fish differ when they live in different depths in the sea? (*relational questions*)	76 (77%)	85 (87%)	
No response	1 (1%)	1 (1%)	

and 70% of the sixth graders answered both items correctly. Although the fourth graders had less experience in scientific inquiry in their science curriculum than the sixth graders, their responses to the items were comparable. Additionally, different types of researchable questions—descriptive (item 1, option A) and relational (item 2, option B)—did not have an impact on either group's ability to identify the questions. However, it may be easier for students who have studied science and other subjects in school for a few years to be able to identify non-researchable questions. If students have learned other subjects, such as social sciences and moral education, that includes posing questions in those fields, this learning experience may help them to recognise whether a question is researchable or not in science. That is, they can see some questions as more suitable ones to ask and answer in other academic and practical fields.

The students identified questions as researchable based on their judgement, and their reasons are elaborated upon in Table 2.2. Their explanations are not sophisticated, perhaps due to their limited epistemic knowledge, but they clearly elaborated on the reasons which were associated with their understanding of the nature of science, personal interest and knowledge, and their view of the subject. The students' understanding of the nature of science, a perspective on the testability of questions through observation and experiment, was used in their judgement. Of the total participants, 33% and 16% of the fourth-grade students and 46% and 29% of the sixth-grade students mentioned this in items 1 and 2, respectively. For both items, sixth-grade students used the concept of testability more frequently than fourth-grade students.

For example, Student 1 (sixth grade, male) mentioned the difference between the two questions in Item 1 from the perspective of testability in the interview. He manifested the use of the elimination method in the interview.

Interviewer: Why did you think that (choice) A in Item 1 is a more suitable question for science learning?

Student 1: As (choice) B changes every day, I do not know if I have learnt about it. For (choice) B, people cannot ask a butterfly directly, therefore I thought of A as I could not find it without asking the butterfly.

Another perspective included in their understanding of the nature of science was objectivity in scientific inquiry that can contribute to sharing knowledge and results in a consensual manner, by avoiding the influence of personal interests and preferences. The concept of objectivity to identify researchable questions was only used in Item 2, which included questions about the appearance of the fish. Students from both

Table 2.2 Students' reasons for categorising questions as researchable

Item	Fourth grade students ($N = 99$) Person (%)	Sixth grade students ($N = 98$) Person (%)	p-value
Item 1. *Butterflies in the cabbage field*			
Testability It could be answered or not through observation and experiment	33 (33%)	45 (46%)	0.21 *n.s*
Objectivity It is dependent on personal preferences	0 (0%)	0 (0%)	
Personal interest and knowledge I know it or I want to know it	39 (39%)	28 (29%)	
Science-like It looks like science or other subjects	14 (14%)	19 (19%)	
Others	13 (13%)	6 (6.1%)	
Item 2. *Fish in the shallow and deep sea*			
Testability It could be answered or not through observation and experiment	16 (16%)	28 (27%)	<0.05
Objectivity Question is dependent on personal preferences	9 (9%)	16 (16%)	
Personal interest and knowledge I know it or I want to know it	35 (35%)	33 (34%)	
Science-like and other subject-like It looks like science or other subjects	20 (20%)	6 (6%)	
Others	19 (19%)	15 (15%)	

grades mentioned that the question is based on personal preferences and not suitable for scientific inquiry. Using this method of elimination, it was assumed that the other question was researchable. For example, Student 2 (fourth grade, male) stated that determining whether the appearance of the fish is good or bad depends on personal preferences.

> Interviewer:Why did you think that (choice) B in Item 2 was more suitable as a question for science learning?
>
> Student 2:That is because I can find out about the colours, shapes, sizes, etc. by researching. Therefore, I thought that I can examine this one (pointing to B). (Choice) A states that the fish looks bad. However, I think that B is the correct choice because the perceptions of people about what looks bad may be different.

Approximately 30% of the students from each grade believed that the questions were researchable, based on the perspective of personal interest and knowledge gained from previous science learning, and wanted to know more about the topics. Student 3 (sixth-grade, female) questioned whether the colour of the fish changed if those living in the shallow sea moved to the deep sea.

Interviewer: Why did you think that (choice) B in Item 2 was more suitable as a question for science learning?

Student 3: That is because I wondered if the colour of the body would get darker, not the looks, and how the colour of fishes would be if the fish living in the shallow sea moved to the deep sea. That is why I chose (choice) B.

Student 4 (fourth-grade, male) identified question A in Item 1 as researchable by reflecting on previous learning experiences as the body structure of insects and their lives were studied in the third grade.

Interviewer: Why did you think that (choice) A in Item 1 was more suitable as a question for science learning?

Student 4: I studied about A when I was in third grade. However, I did not study this question (choice B) when I was in third grade.

The remaining students could not explicitly state their reasons, but they identified that one of the questions was scientific by using their meta-knowledge about different school subjects which they were familiar with, including science. The students mentioned that this question was different from the questions that they were familiar with in science lessons.

Through *chi-square testing*, statistically significant differences in students' reasons for Item 2 were found from the point of view of grade differences ($\chi^2 = 13.30$, $df = 4$, $p < 0.05$). Sixth-grade students answered from the perspective of testability more than fourth-grade students, whereas fourth-grade students preferred to rely on the fitness of their ideas with respect to the subject.

2.3.2 Students' Ability to Pose Researchable Questions

The students' ability to pose researchable questions was explored by analysing their answers to the paper–pencil test. The types of questions based on interrogatives that were extracted from the answers for *heating a solution in the microwave* are shown in Table 2.3.

Through *t-testing*, statistically significant differences were found in the number of questions asked by students in each grade ($t = -5.22$, $df = 190.5$, $p < 0.05$).

Table 2.3 Students' answers as researchable questions

Types of questions based on interrogatives (Examples)	Fourth grade students ($N = 99$) total (mean)	Sixth grade students ($N = 98$) total (mean)	p-value
Better Researchable questions			
How-questions How does salt solution change by heating? How does salt work?	7 (0.07)	22 (0.22)	
What-questions What is the relationship between the weight of water and heating it? What difference is there between the glass cup and the plastic cup?	43 (0.43)	76 (0.78)	
Which-questions Which is easier to heat, water or salt solution? Which can be heated easily, the cup in the centre or the one on the side?	3 (0.03)	7 (0.07)	
Where-questions Where did the salt disappear?	0 (0.00)	1 (0.01)	
Poorer researchable questions			
Closed questions Does the temperature of the solution rise when the cup is heated in the centre? Does the weight of water change its temperature?	30 (0.30)	73 (0.74)	
Non-researchable questions			
Why-questions Why does adding salt raise the temperature? Why is the temperature different when the solutions were heated in the centre or the side?	184 (1.86)	296 (3.02)	
Others Uncompleted questions Declarative statements	40 (0.40)	26 (0.27)	
Total	307 (3.10)	501 (5.11)	<0.05

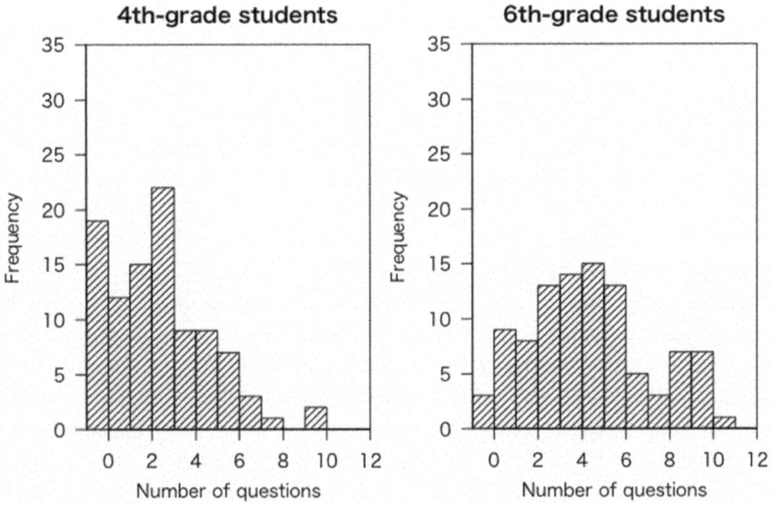

Fig. 2.3 Number of questions asked by 4th- and 6th-grade students

According to the number of students' questions, from which incomplete questions and declarative statements were excluded, the distribution of students from different grades is shown in Fig. 2.3. While the fourth-grade students generated 2.8 questions by mean, the sixth-grade students asked 4.9 questions by mean. In this case, 19% of the fourth-grade students and 3% of the sixth-grade students could not ask any questions. Sixth-grade students asked more questions than fourth-grade students, that is, grade level differences were found in the number of questions asked by the students.

The most popular questions, based on interrogatives, were the *why-questions.* A total of 184 *why-questions* (1.86 per questions student) were asked by fourth-grade students and 296 (3.02 per student) by sixth-grade students. At least one *why-question* was asked by 71% of the fourth-grade students and 85% of the sixth-grade students. This category included questions about the effect of changing variables on the temperature change of the solutions, such as 'Why does adding salt raise the temperature?' and 'Why is the temperature different when solutions are heated in the centre or the side?' These examples could be interpreted to be researchable because they involve two different variables that could be operated if the students can restate the questions without using the interrogative word 'why'. These would be more effective if they also mentioned the relationship between variables. Conversely, some parts of the *why-questions* involved questions about the conditions in the uncontrolled experiment. Other *why-questions* focused on only one variable. For example, Student 5 (sixth grade, female) stated:

Interviewer: (for Item 3) What kind of things did you pay attention to or be aware of when asking the question?

Student 5: Why is this different from here, why is the condition not the same? If this container is used here, why is that container different? I wonder why the conditions are not the same.

These questions cannot be considered researchable. From these examples, it is evident that it is easy for primary school students to ask questions with the interrogative 'why' in science learning. Therefore, they not only ask operational questions, but also intention-seeking questions. This could be the reason for not recommending *why-questions* in scientific inquiry (Alfke 1974).

The second-most popular questions were the *what-questions,* followed by the third-most popular questions, the *closed questions.* Thirty *closed questions* (0.30 per student) were asked by fourth-grade students and 73 (0.74 per student) by sixth-grade students. This category included questions that asked if a specific result was observed, such as, 'Does the temperature of the solution rise when the cup is heated in the centre? and 'Does the weight of water change its temperature?' *Closed questions* are about basic information that can be found in the textbook or are related to some simple observations. Factual questions, as a type of basic information questions, usually require only the recall of information and are often *closed questions* (Chin and Brown 2002). These questions are frequently asked by students in a science class and are considered lower order questions (Eliasson et al. 2017). Despite including the manipulation of variables, a *closed question* is considered a type of confirming question based on the expected results. Therefore, these questions should be formulated with proper interrogatives for further inquiry.

A better form of questions for scientific inquiry is *what-questions.* Forty-three such questions (0.43 per student) were asked by fourth-grade students and 76 (0.78 per student) by sixth-grade students, and 18% of the former group and 29% of the latter asked at least one *what-question.* This category included questions about the relationship between the variables or the effect of a parameter on the temperature change of the solutions, such as 'What is the relationship between the weight of water and heating it?' and 'What is the difference between the glass and the plastic cups?' Other questions were considered more effective were about changes in variables by operation, such as 'How does salt solution change by heating?' In primary science textbooks in Japan that are authorised by the Ministry of Education, Culture, Sports, Science and Technology, a problem-solving sequence is typically preferred as the structure of the textbook. Many questions that have different roles in learning science have been found in science textbooks, especially major questions in the posing-questions phase which are *how-questions*, *closed questions*, and *what-questions*, from the point of view of the usage of interrogatives. *Why-questions* are placed in the perihelial position in science textbooks (Nakayama and Saruta 2015). This preference seemed to be not very different among different textbook companies that have an impact on science teaching because the MEXT has authorized these textbooks. The participating students in this study used one of the science textbooks with such a structure. If a science teacher uses the science textbook as an important teaching resource, their instruction could reflect the structure of it, and students might learn to pose questions for scientific inquiry through such instruction. However, the types of interrogatives that were used by students when they asked questions in this study differed from the types of interrogatives in the science textbooks reported by Nakayama and Saruta (2015).

A limited number of students could pose researchable *what-questions*, and most of them were older students. While some posed researchable questions, they simultaneously asked many unsuitable questions for scientific inquiry, such as intention-seeking questions. These students who asked a mixture of researchable and unsuitable questions were likely in the progressions of developing the ability to pose questions for scientific inquiry. Additionally, some students did not ask researchable questions, but seemed to have a proper understanding of the variables. For example, Student 6 (sixth-grade, male) asked *why-questions* only and mentioned a pair of two different things by comparing unstructured experiments.

Student 6: For all my questions, I thought that by finding (a pair of) two different things, I could ask questions.

This implies that the student understood the variable that could be changed. Student 7 (sixth-grade, male), who asked one *how-question* and four *closed questions*, also explained the importance of the controlled experiment from his learning experience based on understanding the importance of experimental condition control in scientific inquiry in the fifth-grade science curriculum.

Student 7: The amount of water, how to dissolve the salt, and the place to put it is different, and this cannot be compared if it is all different. This is related to (next item) No. 2. When I focus on the amount of salt, I can control everything (in an experiment) except salt. I can also do that in other (conditions).

On the other hand, because the importance of experimental condition control is emphasized in fifth-grade in the Japanese science curriculum, fourth-grade students who had not learned experimental condition control could not mention it. The students need to understand experimental condition control that includes the manipulation of variables by eliminating, substituting, and increasing or decreasing the presence of the variable. In addition, it is also essential for students to acquire epistemic knowledge of asking questions for inquiry and to appropriate scientific discourse, in order to pose researchable questions. The students might have wondering questions and primitive questions that were not yet formalized to researchable questions when observing a phenomenon. Then, they have to translate these ideas into researchable questions in line with a scientific discourse as a social language. Appropriating scientific discourse is not to simply acquire a huge number of technical terms, but also to gain nominal and procedural knowledge in science by participating in social and cultural practices (Rosebery et al. 1992). The ability to talk science is developed by talking to people who have already mastered the language of science. It is learned by using this language to suit one's purpose (Lemke 1990). A science teacher should model the use of scientific language. When a teacher poses questions for scientific inquiry, he/she is recommended to ask *what-questions* and *how-questions*, while avoiding asking *why-questions* as far as the teacher can. In addition, the differences between researchable questions and non-researchable questions in the scientific inquiry and the types of researchable questions, such as *descriptive, relational, and cause-and-effect questions*, should be emphasised in science classrooms.

2.4 Conclusion

Question-driven learning, such as problem-based solving and inquiry-based learning, has long been emphasised in science education. Recently, posing questions as a science practice is emphasised more. This study elaborates on how primary school students identify and ask researchable questions. Most fourth-grade students and sixth-grade students in this study could identify researchable questions. In some cases, they selected descriptive questions as the researchable questions more than intentional questions, and in the other cases, relational questions more than questions related to personal preference. There was no statistical difference in the identification of researchable questions by students' grade level and gender. Different types of researchable questions did not have an impact on the students' performance in this study. The responses to the *paper and pencil test* and interviews suggested that it was easy for students who previously studied science and other subjects in school to identify non-researchable questions. Students identified researchable questions based on their own judgement that was related to their understanding of the nature of science, their personal interests and knowledge, and their view of the subjects. The perspective regarding the testability of questions and objectivity in the scientific inquiry was also seen.

The students' questions based on interrogatives were identified, and it was observed that the most popular types of questions were *why-questions*. While this question category included questions that enquired about the effect of changing variables on the temperature change of the solutions, others were non-researchable and focused on the intention for the uncontrolled experiment. Additionally, *closed questions* were categorised as non-researchable, as they included questions asking whether a specific result would be seen or not. It was determined that *what-* and *how-questions* were better suited for scientific inquiry. Overall, only a limited number of students posed researchable questions, and older students performed better than younger ones. Additionally, some of the students who did not ask the correct type of researchable questions seemed to understand the manipulation of variables and the importance of controlled experiments. This study suggests that students should be taught procedural and epistemic knowledge regarding asking questions for inquiry, including the manipulation of variables by eliminating, substituting, and/or increasing or decreasing the presence of the variable. They need the teachers to model appropriate scientific discourse, including the use of interrogatives. In line with this suggestion, the influence of classroom discourse that includes teacher-student interactions on students' questioning for scientific inquiry could be explored. It remains to be investigated further how the students' understanding of the nature of science improves their posing questions for scientific inquiry and how students' questions drive the scientific inquiry process.

2.5 Note to Our Future Colleagues

Asking questions in education has been essential since ancient times, and its significance can never be dismissed; rather, it should be emphasised further in future education. For teachers, asking questions to students is the simplest and one of the most effective ways to stimulate their thinking and to make them realise where their reasoning may be false. However, teacher questioning often constrains classroom discourse and limits students' thoughts and communication. Teacher education programs for pre-service and in-service teachers that provide insights on instructional scaffoldings in classroom discourse, including questioning, are in greater demand from the perspective of social language use. For students, question-driven learning helps them construct knowledge (Kuhn et al. 2020). Moreover, the important aspects of asking questions are closely related to the ability of critical thinking and creative thinking. Although some research has reported that the number of students asking questions depends on their age and that older students seldom ask questions (Dillon 1988; Osborne and Reigh 2020), this situation must be changed. The development of a learning environment is needed such that not only younger students, but also older students feel free to ask questions in science classrooms and pursue their prospect questions without meaningless interruptions from others. This issue that each student can pursue not only the questions aligned with a science curriculum, but also their own questions would be solved through individualisation of science learning. The development and implementation of students' question-driven curricula could be fulfilled by enriching inquiry-based science education. In this case, how knowledge base for scientific inquiry and scientific ability are developed concurrently through science lessons is a problem to be solved. Providing an opportunity for open inquiry with peers, different age students, parents and mentors that feel free to participate could be a possible solution. We must recognise that scientific practices are embedded in specific cultures and contexts (Douglas and Megan 2014). Some culturally embedded questions asked by students, who are not considered as mainstream, cannot be valued as researchable questions if the teacher, who has an authority in the science class, might not admit them as alternative questions. We are issued not only the problem of how and what ask questions, but the problem of who ask questions and what purpose the person ask a question. By pursuing socially and culturally inclusive science education, a learning community, wherein all students can ask questions critically and creatively, will be actualized.

Acknowledgements This work was supported by the Japan Society for the Promotion of Science (JSPS) Grant-in-Aid for Scientific Research (Grant Number JP19H01680, JP22H01014, JP23K17599).

References

Abdulla AM, Paek SH, Cramond B, Runco MA (2018) Problem finding and creativity: a meta-analytic review. Psychol Aesthet Creat Arts 14(1):3–14. https://doi.org/10.1037/aca0000194
Achieve (2013) Next generation science standards: for states, by states. Retrieved from https://www.nextgenscience.org/. Accessed 10 Jan 2022
Alfke D (1974) Asking operational questions. Sci Child 11:18–19
Allison AW, Shrigley RL (1986) Teaching children to ask operational questions in science. Sci Educ 70(1):73–80. https://doi.org/10.1002/sce.3730700109
Blonder R, Mamlock-Naaman R, Hofstein A (2008) Analyzing inquiry questions of high-school students in a gas chromatography open-ended laboratory experiment. Chem Educ Res Pract 9(3):250–258. https://doi.org/10.1039/B812414K
Chin C (2006) Classroom interaction in science: Teacher questioning and feedback to students' responses. Int J Sci Educ 28(11):1315–1346. https://doi.org/10.1080/09500690600621100
Chin C (2007) Teacher questioning in science classrooms: approaches that stimulate productive thinking. J Res Sci Teach 44(6):815–843. https://doi.org/10.1002/tea.20171
Chin C, Brown DE (2002) Student-generated questions: A meaningful aspect of learning in science. Int J Sci Educ 24(5):521–549. https://doi.org/10.1080/09500690110095249
Chin C, Kayalvizhi G (2002) Posing problems for open investigations: what questions do pupils ask? Res Sci Technol Educ 20(2):269–287. https://doi.org/10.1080/0263514022000030499
Chin C, Osborne J (2008) Students' questions: a potential resource for teaching and learning science. Stud Sci Educ 44(1):1–39. https://doi.org/10.1080/03057260701828101
Cuccio-Schirripa S, Steiner HE (2000) Enhancement and analysis of science question level for middle school students. J Res Sci Teach 37(2):210–224. https://doi.org/10.1002/(SICI)1098-2736(200002)37:2%3c210::AID-TEA7%3e3.0.CO;2-I
Dillon JT (1988) The remedial status of student questioning. J Curric Stud 20(3):197–210. https://doi.org/10.1080/0022027880200301
Dori YJ, Herscovitz O (1999) Question-posing capability as an alternative evaluation method: analysis of an environmental case study. J Res Sci Teach 36(4):411–430. https://doi.org/10.1002/(SICI)1098-2736(199904)36:4%3c411::AID-TEA2%3e3.0.CO;2-E
Douglas LM, Megan B (2014) Who's asking? Native science, western science, and science education. MIT Press, MA
Eliasson N, Karlsson KG, Sørensen H (2017) The role of questions in the science classroom—How girls and boys respond to teachers' questions. Int J Sci Educ 39(4):433–452. https://doi.org/10.1080/09500693.2017.1289420
Hartford F, Good R (1982) Training chemistry students to ask research questions. J Res Sci Teach 19(7):559–570. https://doi.org/10.1002/tea.3660190705
Herranen J, Aksela M (2019) Student-question-based inquiry in science education. Stud Sci Educ 55(1):1–36. https://doi.org/10.1080/03057267.2019.1658059
Hiroshi N, Uchinokura S (2019) Tyuugakusei no kagakutekini tankyuukanounatoi no handan to seisei no jissai: Daigakusei tono hikakuni motoduite [An examination of junior high school students' ability to identify and ask investigable questions: a comparison with University Students]. J Res Sci Educ 60(1):173–184. https://doi.org/10.11639/sjst.18047
Hofstein A, Shore R, Kipnis M (2004) Providing high school chemistry students with opportunities to develop learning skills in an inquiry-type laboratory: a case study. Int J Sci Educ 26(1):47–62. https://doi.org/10.1080/0950069032000070342
Hofstein A, Navon O, Kipnis M, Mamlok-Naaman R (2005) Developing students' ability to ask more and better questions resulting from inquiry-type chemistry laboratories. J Res Sci Teach 42(7):791–806. https://doi.org/10.1002/tea.20072
Keys CW (1998) A study of grade six students generating questions and plans for open-ended science investigations. Res Sci Educ 28:301–316. https://doi.org/10.1007/BF02461565
Krajcik JS, Czerniak C, Czerniak CM, Berger C (1999) Teaching children science: a project-based approach. McGraw-Hill, Boston

Kuhn D, Modrek AS, Sandoval WA (2020) Teaching and learning by questioning. In: Butler LP, Ronfard S, Corriveau KH (eds) The questioning child insights from psychology and education. Cambridge: Cambridge University Press, pp 232–251 https://doi.org/10.1017/9781108553803.012

Lemke JL (1990) Talking science: language, learning and values. Ablex, Norwood, NJ

Lombard FE, Schneider DK (2013) Good student questions in inquiry learning. J Biol Educ 47(3):166–174. https://doi.org/10.1080/00219266.2013.821749

Marbach-Ad G, Sokolove PG (2000) Can undergraduate biology students learn to ask higher level questions? J Res Sci Teach 37(8):1–17. https://doi.org/10.1002/1098-2736(200010)37:8%3c854::AID-TEA6%3e3.0.CO;2-5

Matsubara K (2018) Elementary science education in Japan. In: Lee YJ, Tan J (eds) Primary science education in East Asia: contemporary trends and issues in science education. Springer, Switzerland, pp 49–77. https://doi.org/10.1007/978-3-319-97167-4_3

Mehan H (1979) 'What time is it, Denise?': Asking known information questions in classroom discourse. Theor Pract 18(4):285–294. https://doi.org/10.1080/00405847909542846

Ministry of Education, Culture, Sports, Science and Technology (2008) Shougakkou gakushuu sidou youryou kaisetsu Rika-Hen [Explanation for the course of study for elementary school: Science]. Dainippon Tosho, Tokyo

Ministry of Education, Culture, Sports, Science and Technology (2017a) Shougakkou gakushuu sidou youryou kaisetsu rika-hen [Explanation for the course of study for elementary school: Science]. Gakkou Tosho, Tokyo

Ministry of Education, Culture, Sports, Science and Technology (2017b) Tyugakkou gakushuu sidou youryou kaisetsu rika-hen [Explanation for the course of study for junior high school: Science]. Toyokan shuppan, Tokyo

Nakayama H, Saruta Y (2015) Syuugakko rikakyoukasyo niokeru "Toi" no genjyou to rikazyugyou heno shisa [Trends in questions used in Japanese primary school science textbooks and suggestion for science lessons]. J Res Sci Educ 56(1):173–184. https://doi.org/10.11639/sjst.sp14007

National Research Council (2012) A framework for K-12 science education: practices, crosscutting concepts, and core ideas. The National Academies Press, Washington, DC. https://doi.org/10.17226/13165

National Research Council (2013) Next generation science standards: for states, by states. The National Academies Press, Washington, DC. https://doi.org/10.17226/18290

Oliveira AW (2010) Improving teacher questioning in science inquiry discussions through professional development. J Res Sci Teach 47(4):422–453. https://doi.org/10.1002/tea.20345

Organisation for Economic Cooperation and Development (2015) The ABC of gender equality in education: aptitude, behaviour, confidence. OECD Publishing, Pisa. https://doi.org/10.1787/9789264229945-en

Osborne J, Reigh E (2020) What makes a good question? Towards an epistemic classification. In: Butler LP, Ronfard S, Corriveau KH (eds) The questioning child insights from psychology and education. Cambridge University Press, Cambridge, pp 281–300. https://doi.org/10.1017/9781108553803

Otawa H, Kinoshita H (2014) Tyugakkourika niokeru kagakutekinanouryoku no ikuseinikansuru kisotekikenkyu [Basic research on the instructional method for scientific competencies in lower secondary school science]. J Res Sci Educ 55(3):265–277. https://doi.org/10.11639/sjst.13062

Rönnebeck S, Bernholt S, Ropohl M (2016) Searching for a common ground—A literature review of empirical research on scientific inquiry activities. Stud Sci Educ 52(2):161–197. https://doi.org/10.1080/03057267.2016.1206351

Rosebery AS, Warren B, Conant FR (1992) Appropriating scientific discourse: findings from language minority classrooms. J Learn Sci 2(1):61–94. https://doi.org/10.1207/s15327809jls0201_2

Runco MA (ed) (1994) Problem finding, problem solving, and creativity. Ablex, NJ. https://doi.org/10.1017/9781108553803

Sakamoto M, Yamaguchi E, Murayama I, Nakashin S, Yamamoto T, Muratsu K, Kamiyama S, Inagaki S (2016) Kagakutekinatoi no Seisei wo Shiensuru Rikajyugyou: Genri/housoku ni motoduku toi no rikainityakumokushite [Theory-inspired questions asked by students as an epistemic activity in science: theory-based inquiry and question-centered instruction]. Jpn J Educ Psychol 64(1):105–117. https://doi.org/10.5926/jjep.64.105

Chapter 3
Questioning Patterns in STEM Learning: A Case Study

Niveda Regunathan, Aik-Ling Tan, and Jaime Koh

3.1 Introduction

Questions serve as starting points and anchors for conceptual learning. Cuccio-Schirripa and Steiner (2000) advanced the notion that questioning is a foundational skill at the core of critical thinking, creative thinking and problem-solving. More recently, Osborne and Reigh (2020) posited that questioning by students serves an epistemic function in helping them concretise the relationships between different concepts even as they construct and refine their understanding of disciplinary knowledge. In their comprehensive analysis of the characteristics of good questions, Osborne and Reigh (2020) argued that questioning is "one of the most important epistemic cognitive acts" (p. 281). They analysed dominant analytic classifications of questions used by researchers and argued that existing frameworks lack the capacity to enable researchers to surface the epistemic acts embedded within questions. For instance, the widely popular classification of closed and open questions derived from Bloom's taxonomy (Bloom 1956) does not always provide a consideration of the cognitive demands required to generate the answers since a closed question can be more cognitively demanding than an open question. Osborne and Reigh (2020) cited an example where the response required was to answer questions comparing two scientific processes such as respiration and photosynthesis. While such questions have only one answer (closed), they require students to engage in comparing and contrasting, which are cognitively demanding. Consequently, it was difficult to

---

N. Regunathan
Nanyang Technological University, Singapore, Singapore
e-mail: niveda002@e.ntu.edu.sg

A.-L. Tan (✉) · J. Koh
National Institute of Education, Nanyang Technological University, Singapore, Singapore
e-mail: aikling.tan@nie.edu.sg

J. Koh
e-mail: jaime@nus.edu.sg

Y. S. Ong et al. (eds.), *A Diversity of Pathways through Science Education*,
https://doi.org/10.1007/978-981-97-2607-3_3

determine the reasoning quality required to generate answers to the questions raised. In fact, existing frameworks for classifying questions neither take context into consideration as in the confirmatory and transformative classification nor connect the nature of the questions to the function using Watts et al. (1997) classification of consolidated, exploration and elaboration questions. While each of the question classification schemes could potentially work for specific research questions, they appear limited in their ability to determine the epistemic purposes of questioning during science learning.

Earlier research has shown that questions raised by students are instrumental for their learning as the questions serve as a valuable resource for both teaching and learning science (Chin and Osborne 2008). Questions can be used to enable students to see connections between different scientific entities and concepts, and in turn help them appreciate the epistemic norms in science (for example connecting data to claims). Scientific knowledge is characterized by being empirical, theory laden, and tentative (Lederman et al. 2002) and is acquired through practices such as inquiry using hypothetico-deductive methods (Wenning 2009). As such, scientists ask three types of questions: ontological to ask what exists, causal to question why things happen, and epistemic to find out how we know (National Research Council 2012). Consequently, students should be exposed to and learn to raise ontic, causal, and epistemic questions during science learning. For instance, when studying about photosynthesis, students can better understand the empirical nature of science by asking about the evidence available to conclude that photosynthesis has taken place.

In this study, we set out to understand the prevalence of the use of three types of scientific questions (ontic, causal and epistemic) when students are engaged in integrated STEM problem solving. The intention is to demonstrate how different types of scientific questions are used when students solve problems presented in integrated STEM activities with science as the lead discipline. We also use students' responses to the different types of questions to better appreciate their levels of understanding. As such, the research questions guiding this study are:

1. What are the types of questions (i.e. ontic, causal, or epistemic) raised by teachers and students when students are engaged in an integrated STEM activity? and
2. What are the levels of connections and understanding displayed by students as demonstrated by students' answers to questions raised by their teachers and peers?

3.2 Theoretical Background

3.2.1 *Questions*

In science learning, the process of inquiry typically begins with scientifically-oriented questions that guide the investigations that follow. The vital role that questions play in defining the trajectory of science investigation suggests that learners must (1) learn to identify and ask relevant epistemic questions related to science, and (2) design ways to obtain evidence to answer the questions. Chin (2007) argued that for fruitful and meaningful cognitive engagement for science learning to take place, the kinds of questions and the way questions are asked by teachers matter.

Questioning is an activity that scientists do (Ford 2015) and questioning enables students to connect their knowledge, thinking, arguments, and investigations together (Lai and Law 2013). Questions can also be used to scaffold students' thinking (Chin 2007) as a pedagogical move to guide students as they construct scientific knowledge. A teacher's modelling of different approaches to asking questions can help students develop a deeper understanding of a phenomena and their own ability to ask questions (King and Rosenshine 1993).

Students raising questions for inquiry on their own without modelling is known as open inquiry. Open inquiry is often considered less effective than guided inquiry since the guidance given to students to progress in inquiry is important (Kirschner et al. 2006). Solving 'wicked' problems (Buchanan 1992) is similar to open inquiry since students are often required to ask their own questions to frame the problem. Appropriate scaffolding in the form of questioning can guide students make sense of authentic real-world problem to make their learning experiences more meaningful.

The ability to ask questions is dependent on factors such as students' abilities to identify contradictions, inconsistencies and ill-defined elements that require resolution (Huang et al. 2017). As such, examining question types and when specific questions are asked during science learning or STEM problem solving enables teachers to assess students' understanding of both disciplinary conceptual understanding. Further, questions asked, and the corresponding students' responses provides insights into inconsistencies and possible contradictions in arguments and explanations during problem framing. Further, understanding the context of questions in relation to a learning activity and the types of questions asked are likely to facilitate teachers to prompt students to ask more questions exhibiting higher-order thinking (Marbach-Ad and Sokolove 2000).

The classification of questions is a widely researched topic. Mortimer and Scott (2003) grouped questions by how teachers functionally use them—Shaping ideas, selecting ideas, marking key ideas, sharing ideas, checking students' ideas, and reviewing. Beyond the pedagogical functions played by questions as highlighted by Mortimer and Scott, researchers applied Bloom's taxonomy in various ways to illumine the cognitive functions of questions. Questioning frameworks such as open-closed questions (Anderson et al. 2001), lower-/higher-order questions (Barden 1995), input-level/processing level/output level questions (Shepardson and Pizzini 1991), and consolidated/exploration/elaboration questions (Watts et al. 1997) have been widely used by researchers. Osborne and Reigh (2020) argued that existing ways of classifying questions fail to convincingly consider the aspects of cognitive demands required to answer the questions and do not show how students connect with the epistemic work of constructing knowledge. In science learning specifically, they argue for a need to consider three categories of questions for students to connect epistemically with science knowledge construction. Firstly, students and teachers can engage with ontological questions where they seek to understand what exists.

Osborne and Reigh (2020) identified two types of ontological questions: basic (simple ontic) that include questions such as *What is the symbol for carbon?,* or key (complex ontic) with questions consisting of two or more concepts (for example, *what is the difference between mass and weight?).* While Osborne and Reigh (2020) used the term "key ontic" questions to denote questions involving two or more concepts, we have termed this group of questions "complex ontic" questions to contrast with the category of "simple ontic" questions. The second type of questions are causal, which are meaningful for constructing scientific concepts. Causal questions establish why something happens and how a specific action could result in a reaction. Examples include *Why is the sky dark at night?* Or *What will happen when a pebble drops into a basin of water?* Finally, the last type of questions that Osborne and Reigh (2020) proposed are fundamental or basic epistemic questions. These questions help us to understand how we know what we know. Examples of such questions include *What is the evidence suggesting that oxygen is given off during photosynthesis?* Or *Can I know why you claim that A is heavier than B?* Given that the intention of this research is to find out how students are engaged with the sense-making process of increasing crop yield through urban farming methods, we adopted Osborne and Reigh's (2020) three categories of questions to uncover what and how questions are used in the sense-making process (Table 3.1).

Questions, together with their corresponding answers, can serve as indicators of students' levels of understanding of a disciplinary concept or a problem scenario. Analysing questions and the answers provided in the context of the activity allow

Table 3.1 Question categories (based on Osborne and Reigh 2020)

Type	Sub-type	Description	Examples
Ontic	Simple	Simple questions about observable entities	• What's respiration? • What's hydroponics? • What are the usual maintenance costs?
	Complex (similar to key questions highlighted in Osborne and Reigh (2020))	Complex questions that combine two or more concepts and seek to understand what exists in relation to a concept	• What are some problems that you can possibly foresee? • What is the technology involved in vertical farming? • What is the difference between these countries and Singapore?
Causal		Questions that establish why something happens and how a specific action could result in a reaction	• How do you water the plants? (As per the context, the question required an explanation of the watering mechanism featured in the prototype) • Why would the triangle shape have more sunlight?
Epistemic		Questions seeking justifications and help to understand how we know what we know	• Where do you get your information from? • Why is this a worthy investment? • Can I know why?

researchers to better recognize the students' levels of understanding of specific phenomena. One such method of interrelating questions, answers, and levels of understanding is the SOLO taxonomy (Biggs and Collis 1982). The SOLO taxonomy consists of five stages of understanding starting with pre-structural where one is not confident of any aspect of the topic and will not be able to raise any relevant questions. The next stage is the uni-structural stage where the learner knows isolated bits of information about a topic and can ask questions related to identification or naming, very much like simple ontic questions. The other three stages are multi-structural, relational, and extended abstract. In these three stages, learners can identify several pieces of information related to the topic (multi-structural) and connect them (relational) and, if they are able to apply the knowledge to other topics, they have reached the stage of extended abstract. In this chapter, we applied the SOLO taxonomy stages and inferred from the students' responses to questions asked so as to understand their levels of understanding. Table 3.2 provides some examples of each level.

3.2.2 Nature of Questions in STEM Classrooms

In a STEM classroom where engineering design is infused along with scientific inquiry, students and teachers learn at the intersections of different disciplines (Brand 2020). Hence, it is possible that the types of questions asked during integrated STEM problem solving can differ from those asked in a typical science lesson. With a focus on design and practical problem solving in engineering, we need to better understand how to use design projects to support students' meaningful learning of STEM concepts (Crismond and Adams 2012). In an integrated STEM classroom where students are engaged with design problem-solving, learning goals include helping learners to transit from problem solving to problem framing, from unfocused troubleshooting to diagnostic troubleshooting, from a state of idea scarcity to idea fluency, and from ignoring analysis of benefits and trade-offs to balancing benefits and trade-offs in their decision making (Crismond and Adams 2012).

The infusion of engineering ideas into science learning naturally shifts learning discourse from being uniquely focused on disciplinary conceptual knowledge to including more practical knowledge. Practical knowledge describes commonsense knowledge that are related to solving issues in the immediate environment where people live, for example, using a door stop to prevent the door from slamming shut when the wind is too strong. Integrated STEM learning creates a need for a balance between scientific and practical/common-sense inquiry, a theoretical dichotomy described by Dewey (1938). The means by which teachers and students balance the questions and discussions to achieve the dual outcome of problem solving and learning of scientific knowledge require more attention.

Current research into questions and questioning in integrated STEM classrooms is scarce. Tagnin and Riodain (2021) found that the use of question-focused instructional approaches during integrated STEM lessons lead to greater usage of

Table 3.2 SOLO taxonomy levels (Biggs and Collis 1982)

Level of understanding	Description	Example of student's response
Pre-structural	Student is unaware of any information related to a specific idea	• I don't know
Uni-structural	Student is aware of one or two isolated pieces of information that constitute an idea	• Water tank will hold water, for water to go on, and then the plants will be on each green platform • And there's also this little timing box over here which is to program, at a certain amount of time, the water will sprinkle around for a set amount of time
Multi-structural	Student is aware of different pieces of information within an idea but is unable to link them	• Carbon dioxide machine, so it will give out carbon dioxide so that the plants can grow better. And this thermostat, it controls the temperature of the whole place • Then got solar panels here then this thing will be transparent to light. For the sunlight to pass through also. So the solar panel will collect electricity also

(continued)

Table 3.2 (continued)

Level of understanding	Description	Example of student's response
Relational	Student can link together pieces of information constituting an idea	• This is the aluminium foil to reflect the light to the plant. The stand is to hold the kalian so that the water can flow in and then the kalian will get water. The base is to hold the plant firmly so they do not topple • In the middle of the structure, there is a Ferris wheel basically, which rotates. And each basket has a kailan. And so basically, one basket can ensure about 15 min of sunlight, so it will receive equal amounts of light. Each basket will have a watering system made up of sprinklers, that waters the plant across. There's also a solar system which stores energy to power the UV light if there is not—To turn on when it's night, so that it's more efficient, so that it can be harvested at an earlier date. So, when there's no light, you can use the UV light, and this will help the kailan to be able to go through photosynthesis. And this will make it easier for us to harvest it at an earlier date, and also make it so that there is lesser space—We don't need to use as much space—we can use as less space as possible. This is our prototype
Extended Abstract	Student can transfer the knowledge of the topic he/she was working on to other topics and is able to see how they link together	• The solar panels are, yes we will need to check on them once in a while, and it may, not that often, but it also may pose a problem because like what happens if it rains? You know, so it has a bit of a problem in terms of weather aspects • And also, the cost of—In total the cost of the resources will be reduced because everything will be reused. And the space right, we are thinking of building it a few floors of the same model, we will build a few floors so we will be saving space

content–Language integrated learning approaches that enabled learners to participate more meaningfully in classroom discourse. Specifically, Tagnin and Riodain (2021) found that teachers' strategic use of questions serves to promote understanding of scientific concepts and use of scientific language. They recommend that STEM teachers should not request and focus only on one-word answers and thus oversimplify the linguistic demands of students by as this could lead to generalisation of the content learnt. In yet another earlier study on questions asked in inquiry-oriented lessons, Bielik and Yarden (2016) found that students' ability to ask meaningful research questions in biotechnology is enhanced through student-centred, dialogic, and interactive teaching strategies. Our current understanding about the types of questions and how these questions shape learning in STEM can be deepened given the importance of questions in shaping conceptual learning and making connections between concepts and problem-solving.

This lack of research examining the types and nature of questions asked during integrated STEM problem solving motivated this study. We aim to describe the types of questions asked and the purposes these questions serve when students are engaged in integrated STEM problem-solving. We hope that the descriptions about questions asked will serve as the starting point for more research on questions in integrated STEM learning.

3.3 Methods

3.3.1 Participants

A total of 214 Grade 8 students from two schools in Singapore participated in this study. The students came from ten different classes and each class is taught by a different teacher. On average, there were between 25 and 30 students per class. Five classes were assigned to the problem-centric teaching approach on the topic of urban farming (N = 105) and the other five classes were assigned the solution-centric teaching approach on the same topic (N = 109).

3.3.2 Lessons

A total of 36 lessons were conducted across all classes and teaching approaches. Each group received at least one activity booklet to assist with each approach's tasks. The students worked on the tasks in groups of four to five. For most of the students, this was probably their first experience with formal integrated STEM activities. For this chapter, we analysed the final lesson (about an hour long each) for all of the 10 classes as the students presented their prototype to the class. The video recordings of these lessons were chosen for analysis as they were where the questions and answers were most audible.

3.3.2.1 Topic and Teaching Approaches

Using the S-T-E-M Quartet instructional framework (Tan et al. 2019) as the base, the urban farming activity was designed to create a learning experience for students to acquire knowledge and skills from different STEM disciplines before solving novel problems or improving existing solutions. The activities were developed by the second and third authors together with their colleagues from the larger research team. Before the lessons, the second author spent three hours with the teachers to familiarize them with the materials. Teachers in both approaches first established to their students the context of limited land space and the problem of food sustainability in Singapore before introducing the tasks. Depending on the time each school allocated for the activity, all tasks (for both approaches) were completed within three to six hours. There were two forms of centricity in planning the activity—Problem-centric and solution-centric (Teo et al. 2021).

Problem-centric—The problem-centric activity required students to design a farm in an urban landscape to produce vegetables to support the local community while considering constraints such as the optimization of land area and manpower as well as the farm's cost-effectiveness. The students engaged in activities to research and learn about different urban farming methods before designing and building a physical prototype of their proposed model. They then presented their model and evaluated their peers' models during a sharing session.

As the problem presented was open-ended, our initial hypothesis was that students were likely to engage in discussions related to framing the problem, identifying its boundaries and proposing possible solutions. Hence, we expected a high level of variability in the models proposed as possible solutions to the problem. Consequently, the questions asked were expected to be more generative and related to understanding the various aspects of the problem (epistemic) and the cause-and-effect (causal) of the various solutions suggested.

Solution-centric—The solution-centric activity required students to develop an improved version of a vertical farming system while considering the same set of constraints as the problem-centric approach. Students were given examples of existing vertical farming systems and were required to describe how the systems work together with the strengths and weaknesses of each system. Upon identifying the limitations of existing systems, students developed an improved version of the vertical farming system that could produce food locally in their school community garden. Students engaged in group discussions related to understanding how the existing vertical farming systems work and explored areas that could be improved. Subsequently, they designed and built a model of their new farming system and presented their solution to the rest of the class.

Although a general scenario was presented, the students were provided with existing solutions to develop further. As such, we hypothesized that the students would have a better understanding of the existing strategies and effectively apply this knowledge to present viable solutions. Hence, the questions asked could possibly centre around evaluation of the various components (ontic) of existing structures as compared with new solutions proposed.

The fundamental difference between the problem-centric activity and the solution-centric activity lies in the starting point. The starting point for the problem-centric activity was the presentation of a problem without a hint of a solution while for the solution-centric activity, the starting point was the presentation of existing solutions.

3.3.3 Data Collection and Analysis

All classroom lessons were video recorded, and the videos were transcribed prior to data analysis. A turn of talk (an utterance) was used as the unit of analysis to identify the questions raised by teachers and students in the lessons. We did not distinguish questions asked by teachers and students because during presentation of ideas at the end of the activity, the teachers and students were equally empowered to raise questions. We first employed deductive coding by adopting Osborne and Reigh's (2020) method of classifying questions into types and sub-types as described in Table 3.1.

In the second stage of our analysis, we examined students' responses to questions asked for levels of understanding of specific scientific phenomena. Each response to questions asked was re-examined and the understanding was deductively coded using the categories described in the SOLO taxonomy (Biggs and Collis 1982). Refer to Table 3.2 for a description for characteristics of each level.

To ensure the reliability of the codes, the transcripts were coded by a first coder (first and second authors). The first and second authors served as the first coder each for half the number of total transcripts. Once a transcript had been coded once, it was sent to the second coder (third author). The second coder randomly selected 10% of the codes to code independently. The second coder is an experienced coder who is familiar with the data set. After this, the codes from the second coder were sent back to the original first coder to check for similarities. In cases where the first and second coders differed, discussions on these differences took place prior to re-coding based on the discussions. This process was carried out until we achieved an inter-rater uniformity of 90% in terms of simple agreement.

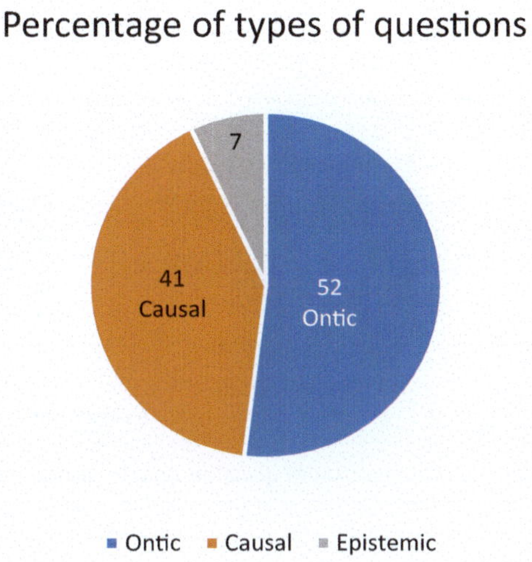

Fig. 3.1 Question types asked across all analysed lessons (both instructional approaches)

3.4 Results and Discussions

3.4.1 Question-Types for Both Approaches

We identified a total of 574 questions asked by students and teachers in the 10 h of lessons. The number of questions asked was not evenly distributed across the 10 h. There were more questions asked in some lessons and fewer in others. Figure 3.1 shows the percentage distribution of question types expressed as a percentage of the total number of questions identified in the 10 h. Regardless of teaching approaches, ontic questions were asked the most frequently (52%), and epistemic questions were asked least frequently (7%) for both teaching approaches.

3.4.2 Comparing Question-Types for Problem-Centric and Solution-Centric Approaches

For the next level of analysis, we examined each question-type to determine if there was a significant difference between the two instructional approaches. Figure 3.2 shows percentage frequency of the question-type from the two instructional approaches.

Corresponding to the overall trend for question-types, for both problem-centric and solution-centric approaches, ontic questions were asked most frequently. Epistemic questions were least frequent in both approaches. Causal questions were equally frequent for both teaching approaches.

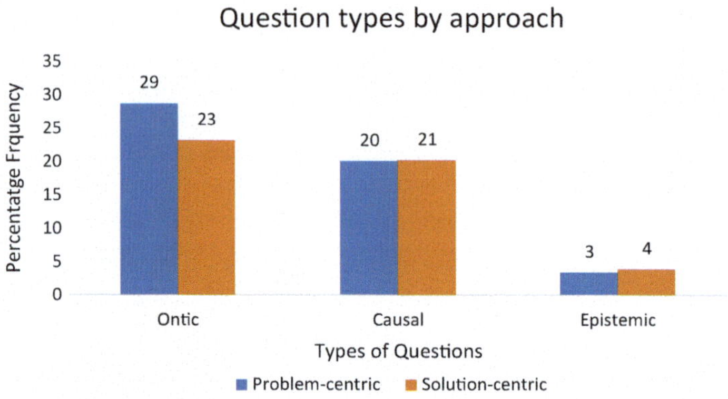

Fig. 3.2 Comparison of question-type by instructional approach

3.4.2.1 Ontic Questions

We observed that there was a numerical difference in the type of ontic questions (simple and complex ontic) asked between the problem-centric and solution-centric approaches. In total, there were 6% more ontic questions asked in the problem-centric approach compared with solution-centric approach (Fig. 3.3). For the problem-centric approach, complex ontic questions were 3% more frequent compared to simple ontic questions. For the solution-centric approach, 7% more simple ontic questions were reported compared to complex ontic questions. A chi-square of independence was performed to examine the relation between the two approaches and the type of ontic questions. The relation between these variables was not significant, X^2 (1, $N = 109$) $= 2.54$, $p = 0.05$. There is no statistically significant difference between the number of simple ontic questions and complex ontic questions asked and the instructional approaches.

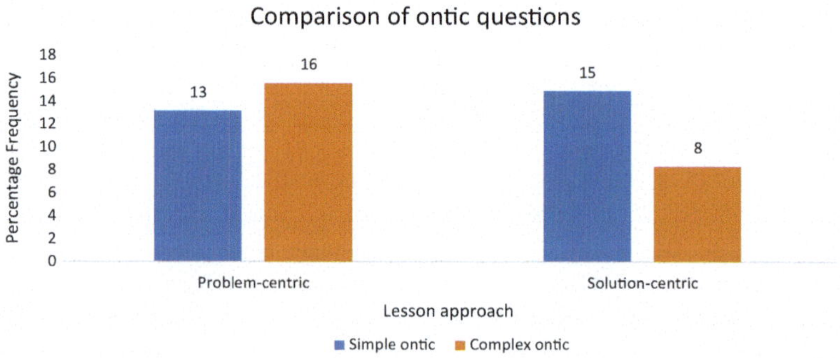

Fig. 3.3 Comparison of number of simple and complex ontic questions in problem-centric and solution-centric teaching approaches

Simple ontic questions can be answered with knowledge from a single concept. For both instructional approaches, the solutions that students presented related to hydroponic farming systems, solar powered greenhouses, and features such as water slides and spiral structures in their prototypes. The simple ontic questions asked were focused on aspects of factors related to photosynthesis, process of respiration and dimensions of various proposed structures. The complex ontic questions related concepts such as the process of photosynthesis and how the proposed structures support these processes (for example, "how to control the water go to like which plant"). The students from the solution-centric approach worked to modify existing vertical farming structures. As such, their solutions were less varied. This could have led to more simple ontic questions where students' questions revolved around what materials the proposed frames would be made of, or whether electricity was required. On the other hand, the complexities and greater variations of solutions suggested in the problem-centric approach could have resulted in students asking more complex ontic questions to clarify the relationship between the photosynthetic process and the proposed farming structures or how the various components of the proposed farming structures are linked to one another.

3.4.2.2 Causal Questions

Questions that established why something happened and how a specific action could result in a reaction were labelled as causal questions. For example, to better understand the farming structures presented, students asked questions such as, "if you make it narrow, you are actually limiting what?" and "Ok, if it rains very heavily, how will it be affected?". In these instances, we observed that causal questions were raised to prompt more critical thinking of the proposed idea. In fact, when these questions were asked, the students presenting their ideas usually responded with longer and more detailed explanations of their ideas. Again, we found no difference in the number of causal questions asked in the problem- and solution-centric approaches. This could be attributed to the fact that for both approaches, prototypes were presented, and causal questions were asked to enhance understanding of how the prototype works under different conditions. Across the lessons, slightly fewer causal questions than ontic questions were asked.

3.4.2.3 Epistemic Questions

Epistemic questions seek justifications and help us understand how we know what we know. They are generally considered important for helping learners understand how evidence that was obtained was used as justification for decisions made. Examples of epistemic questions include "…how are you going to consider this to accelerate the growth of Kai Lan?" and "How do you know that it will give you 3.3. tonnes?" We noted few such higher-order questions in both the problem- and solution-centric approaches. Our results align with Osborne and Reigh's (2020) observations that there are few instances of epistemic questions used in classrooms and hence there is a need to deliberately channel more deliberate efforts to asking epistemic questions in class so as to help learners better appreciate the relationship between evidence observed and claims made.

Generally, the types of questions asked in a class depend on factors such as the knowledge that students have of the phenomena and the context in which the information is presented (Herranen and Aksela 2019). In cases where students are highly knowledgeable about the presented problem and its solution, it is more likely they will raise more epistemic questions. However, if the information presented is sparse and relates only to one aspect of the problem, the accompanying questions tend to be more ontic and causal in nature. While there are no statistically significant differences between simple and complex ontic question, the higher instances of complex ontic questions in the problem-centric approach suggest that the open-endedness of the problem resulted in the need for students to consider more varied factors in order to design the solutions. They needed to understand the physical and concrete presences of each part of the structures in the prototype and the functions of these parts. The problem-centric group (designing an urban farming system) had greater variety to the design of their solutions compared to the solution-centric group (modifications of existing vertical farming systems) resulting in more clarifications about novel prototype designs.

Research showed that cognitive demands can vary depending on the types of questions asked. Erdogan et al. (2017) discovered that questions asked during by teachers and students in 5th, 6th, and 7th grade classrooms in Turkey were typically related to remembering and understanding. Eliasson et al. (2017) found that 87% of questions raised in science lessons were closed questions that were lower order in nature and did not promote the development of higher cognitive thinking among learners. Our data shows similar questioning patterns with a higher frequency of simple ontic questions and a lower frequency of epistemic questions. These studies showed that asking closed-ended questions did little to guide students' thinking processes. Rather, Chin and Osborne (2010) observed that students require teachers' guidance to ask questions requiring higher-order thinking. Hence, teachers can model for students how to ask higher-order thinking questions.

3.4.3 Levels of Understanding Across Both Teaching Approaches

We identified a total of 165 utterances (defined by a single turn of talk) in response to the 574 questions asked in both types of classrooms. These utterances were made by students and reflected various levels of understanding as described by the SOLO taxonomy. 90% of the 165 utterances were categorised as uni-structural, multi-structural or relational. Figure 3.4 shows the distribution in the levels of understanding across all the lessons.

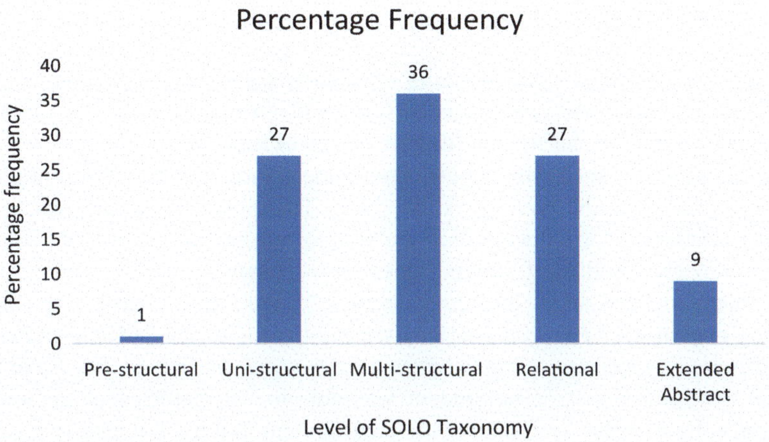

Fig. 3.4 Levels of understanding observed across both teaching approaches

3.4.3.1 Pre-structural Level

Utterances identified as showing pre-structural, the lowest level of understanding, were observed least frequently across both teaching approaches. This could be attributed to two reasons—The first reason being that the students did not often voice their lack of knowledge, and the second being that students were given time to search for information online during the activity and were able to apply their knowledge in the building of the prototype.

3.4.3.2 Uni-Structural Level

Uni-structural level of understanding occurred as often as relational level of understanding across both teaching approaches. Uni-structural levels of understanding were identified when students responded using mono-syllabic answers without elaboration to questions asked. For instance, in response to a question of "*How do you harvest the kailan*?" (causal question as it was asked with reference to a specific structure of the design), a student replied, "*just take out the plant la!*" There was no evidence of elaboration of how the structures could enhance the ease of harvesting the kailan. In yet another example, in response to a question of "why do you need the barrier?" (causal question), the student answered, "*for the plants to have like more space to grow*". There was no evidence of connection to how the barrier provided more space for the plant to grow.

3.4.3.3 Multi-structural Level

Multi-structural level of understanding was most frequent across both teaching approaches. In both approaches, students were given opportunities to gather and internalise information during the different stages of research, discussion, building, and presentation. As such, students showed understanding of the different aspects of urban farming and the prototypes they built. There were fewer instances where students made relational connections. For example, when a teacher from the solution-centric approach posed the following question—"What are the abiotic and biotic factors that you have taken into consideration?", the student answered with the following response—*"For sunlight, we used the light with the LED lamps because this hydroponic system is not super affected by the environment. So like it can grow in harsh conditions also."* The student was able to list factors such as sunlight, LED lamps, hydroponic system, and the environment, but did not clearly articulate the links between these different components to answer the complex ontic question asked. Although the student could have known the links, failure to explicitly mention the links caused this utterance to be classed as understanding at the multi-structural level instead of relational level. An example of a response clearly articulating the links would be *"Light is an abiotic factor of photosynthesis. Instead of sunlight, we used LED lamps as the plants in the hydroponic system will be able to photosynthesize even when there is no natural light."*

3.4.3.4 Relational Level

Across both teaching approaches, students showed a relational level of understanding of the idea of vertical farming and photosynthesis. At the relational level of understanding, students can link together different pieces of information related to a specific idea. This level of understanding is important as it suggest that students were able to make connections between different ideas during situations of complex problem-solving and decision making. For instance, when the teacher asked if the circular shape of the farming structure was to ensure that every plant obtained the same amount of water, the student replied that that was the case and elaborated that "*if there's a drought, then the plants can still get some water.*" The student was able to relate the idea of conditions of drought, plants' need for water to the design of the farm.

3.4.3.5 Extended Abstract Level

We observed infrequent instances of utterances at the extended abstract level, which is found at the highest end of the SOLO taxonomy spectrum. At this level, students showed that they were able to make connections between photosynthesis and farming systems, and wider concepts of sustainability and food security. For example, in explaining their farming system, one student described the prototype *as "so how it works is that it's much more bigger and taller, so it acts as a windmill. So the windmills are by kinetic energy—It converts kinetic energy into electrical energy, so that it will have a reusable source of energy.*" In this instance, the student related the farming system to the idea of sustainable energy, which is a concept that is not primarily related to photosynthesis but is relevant to the idea of sustainable farming systems. In another similar example, a student presented the following, "*You know those dams, and then the dams will go like down the water, then there's this kind of hydro thingy. That spins when it touches water. Spins then generates electricity. So right, when the water falls down, then it will spin. The water will—Fan-blade thing, then it will spin, and then it generates electricity. So, it saves electricity, it gets electricity and also saves water.*" These students showed they were able to connect the concept of plant nutrition (a biological topic) to the transfer of energy and electricity (a physical science topic). This ability to connect relevant ideas beyond that which is currently being discussed requires knowledge and understanding of the relationship between the concepts and involves higher-order thinking skills.

3.4.4 Comparing Levels of Understanding in Problem- and Solution-Centric Approaches

Figure 3.5 shows a comparison between the levels of understanding for problem- and solution-centric approaches. The percentage frequencies were based off the total number of utterance (N = 165). It can be seen here that more instances of students' utterances were identified in the solution-centric classrooms compared with problem-centric classrooms. In this analysis, we did not take the length of an utterance into consideration. A chi-square of independence was performed to examine the relationship between the two approaches and the levels of understanding of the vertical farming systems exhibited by the students. The relation between these variables was statistically significant, X^2 (4, $N = 165$) = 14.68, $p = 0.05$. The instructional approaches affected the levels of understanding of issues related to farming systems that were presented to the students.

Comparing problem- and solution-centric classes, students in solution-centric classes showed higher percentage of uni-structural, multi-structural, and relational level of understanding. Only for the extended abstract level of understanding were there more students in problem-centric classes compared with solution-centric classes (7% compared to 2%) as in Fig. 3.5. Excerpt 1 is an example of a student exhibiting extended abstract level of understanding.

Excerpt 1. Example of extended abstract level of understanding.

Student: *This is just part of the building. It's not the whole thing. And some parts of it are very sophisticated, so we didn't really have time. So, first off, we have these panels right, that are what you call solar (inaudible). They are for conserving energy for both the day and night. So we can reuse energy. Because we are going to put this*

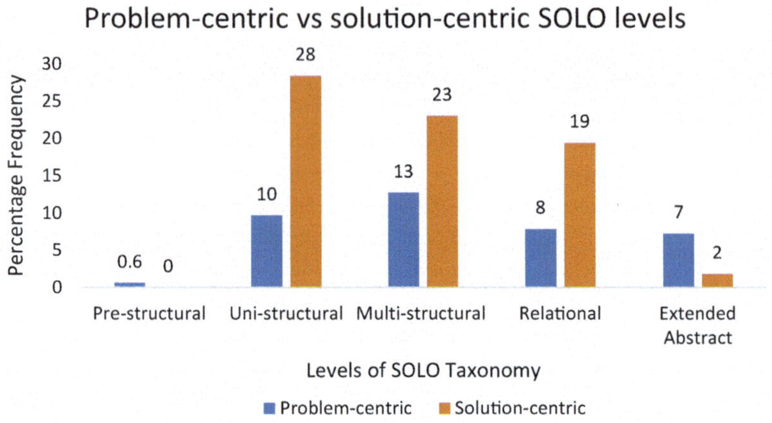

Fig. 3.5 Comparison of SOLO levels in problem-centric and solution-centric classrooms

in an enclosed room, therefore we can trap the heat and we will have a light source inside, using energy that comes from the panels. Then it will have these sprinklers that will use rainwater. When they sprinkle right, the excess water from the soil will be used again to water the plants. Therefore it will save resources. So this is a very vertical—like we will put a few floors of it, so therefore these are all like inaudible. Some advantages I will think about, is most of the things are using machines that are automated by time—timing automated. So at a specific time they will activate. Therefore, reducing manpower.

At the extended abstract level of understanding, a student is able to transfer the knowledge of the topic he/she was working on to other topics and is able to see how they link together. In excerpt 1, the student presented information such as the solar panels and sprinklers and subsequently linked them to how they affect photosynthesis. Further, the student made connections between the concept of photosynthesis and other ideas such as energy, conservation of energy and sustainability. Other utterances at extended abstract level exhibited similar patterns. From these observations, we inferred that the openness of the problem presented students with opportunities to explore beyond vertical farming systems. This wider scope also allowed students to explore divergently and gather more facts and evidence to support their decisions and designs.

While the problem-centric approach showed more instances of responses at the extended abstract level (19%), the solution-centric approach showed higher frequencies of uni-structural (28%), multi-structural (37%) and relational (31%) levels of understanding. Instances of utterances at the uni-structural level showed a common pattern comprising back-and-forth exchanges between a teacher and a student. Excerpt 2 is an example of such an exchange between a teacher and a student. The student required constant facilitation by the teacher to answer the question fully. As seen in excerpt 2, constant probing by the teacher was needed to move the discussion forward as the student was uncertain how connections could be made.

As such, this excerpt was taken as a single utterance of unj-structural level of understanding as the statements were all made by one student in a single conversation. This form of exchange occurred more frequently in classes using the solution-centric approach. While there were many instances of probing observed in the solution-centric approach, there were also equally many instances of relational level of understanding. One possible reason why there were as many instances of relational level of understanding could be that the students in the solution-centric activity were all working on a narrowly focused activity to improve a vertical farming system and as a result were all equally familiar with the activity. Consequently, all the students were given opportunities to think through the features, strengths, and limitations of vertical farming systems, and in turn, teachers were better able to draw links between concepts when questioning their students and when critiquing the ideas presented by others.

Excerpt 2. Example where uni-structural level of understanding is exhibited.

Teacher:	What's the context? What do you think we need to talk about in context?
Student:	Finding microplastics in sand.
Teacher:	Finding microplastic from sand. The sand is where?
Student:	At the beach.
Teacher:	At the beach is at where?
Student:	East coast.
Teacher:	At East Coast Park is at where?
Student:	Singapore.
Teacher:	Singapore. So what I'm trying to tell you is that there are (inaudible) smaller context and larger context.

3.5 Conclusion and Implications

In this chapter, we set out to describe the types of questions that teachers and students asked as they solved integrated STEM problems. With reference to the first research question "What are the types of questions raised by teachers and students when students are engaged in an integrated STEM activity?", we found that ontic questions, both simple and complex, were most common, compared with causal and epistemic questions. Epistemic questions were least commonly asked in class. The popularity of ontic questions suggest that these questions are needful to help learners establish fundamentally accurate disciplinary knowledge as they engage in practical problem solving. The role of basic disciplinary knowledge has been described by various scholars (for example Cuthbert 2021, p. 20) as a means for learners to engage cognitively with "universalizing human thought and experience" so that knowledge can shift from personal opinion and beliefs, to ground knowledge in more abstract conceptual systems. Our findings show similarities with that of Erdogan et al. (2017) and Eliasson et al. (2017) who showed basic levels of questions are the most commonly asked in class. The challenge lies in lifting the cognitive levels of the questioning to a higher level and a more epistemic purpose. The value of asking epistemic questions lies in enabling students to better understand the basis for believing or trusting the knowledge they encounter as they would understand how the knowledge is produced and justified. While we did not find any statistically significant differences in the types of questions raised in problem-centric or solution-centric approaches, we inferred that regardless of the teaching approach, students and teachers asked a variety of questions, particularly ontic and causal, as they sought to understand the problem/solution presented. One important insight that we gleaned from categorizing the questions raised is the need to consider the context in which the question is asked. Unlike in a science classroom where questions asked are usually more straightforward as they pertain to scientific investigations or how explanations are related to observations, questions in integrated STEM problem-solving are more complex. If the context of the questions is not considered, it is likely that the questions would be misunderstood by students or mis-categorised by researchers.

We used the students' responses to the questions posed by teachers or peers to infer their levels of understanding related to the issue of farming systems. With reference to the second research question, *"What are the levels of connections and understanding displayed by students' responses to questions raised by their teachers and peers?"*, we found that the type of instructional approach affects the level of understanding exhibited by students. The most common level of understanding shown by participating students were at the multi-structural level of understanding where students described or listed the various entities involved in farming systems although they were less able to make meaningful connections between these entities. Relational levels of understanding were also observed although at a lower frequency. Extended-abstract levels of understanding was even less frequently exhibited. Based on the frequency of different question-types, we inferred that for both teaching approaches, a similar sense-making process in critiquing a plausible solution/prototype to a STEM issue was involved. While the frequency of the questions asked may be similar, the levels of understanding exhibited by the students differed. In a more open-ended problem-centric environment, students exhibited extended-abstract understandings more frequently compared to solution-centric classrooms. The more open-ended nature of the problem afforded more opportunities for students to explore concepts that are not directly related.

Our findings suggest that engagement in an open-ended problem-centric activity or a closed solution-centric activity does not affect the frequency of the three different types of questions asked. Designing STEM solutions require students to find out the various components or physical entity related to the specific phenomena and hence students ask ontic questions. The understanding of the facts about the entities involved in the phenomena usually leads students to establish the relationship between the entities by asking causal questions. Our observations suggest that epistemic questions are only asked when students are doubtful of the ideas presented to them and hence seek to validate the evidence available.

Based on our results, we would like to suggest two ways to use integrated STEM activities to achieve the goal of science learning and problem-solving. Firstly, teachers can make deliberate efforts to incorporate more epistemic questions to complement the ontic and causal questions asked. Examples of epistemic questions that teachers could ask during integrated STEM activities are "why did you design it this way?", "Why do you think it is a good idea to increase the size?" Doing so will help students better understand the role of empirical evidence in supporting ideas or solutions presented. Secondly, teachers can consider presenting integrated STEM activities as open-ended problems to students so as to encourage them to explore all possible related disciplinary concepts and cross-disciplinary connections as they devise solutions. Making connections between disciplinary concepts is an important outcome for integrated STEM learning (Honey et al. 2014).

3.6 Limitations

This study was undertaken as part of a larger study comparing students' learning experiences during problem-centric and solution-centric STEM activities. The analysis presented here is from the video recordings of the presentations of ideas to the class. However, sessions with group discussions were not analysed. It is possible the two approaches could elicit different distributions of question types during group discussions. Secondly, we acknowledge that the number of questions analysed could have been suppressed due to poor audio recording quality as the students were wearing surgical masks during the lessons as part of the pandemic safe management measures. However, as all students and instructors were wearing masks, any suppression of identifying questions should have affected both approaches equally.

3.7 Note to Our Future Colleagues

The year is 2050, and integrated STEM teaching and learning is the default classroom practice. Your students and you are constantly working on creating solutions for real world problems. Students are learning disciplinary knowledge through asking, doing, making, and refining. Science, mathematics, and design are now taught in an integrated manner. Students are wondering how students in 2022 learned since it makes little sense to learn things in silos. It is intriguing how students in 2022 could make connections across different disciplines when they learned them separately. Teacher education and professional development in 2050 has also been transformed. Courses are offered to focus on designing interdisciplinary activities and how enactment of interdisciplinary problem solving. Teachers work in teams where everyone brings this expert disciplinary knowledge to design meaningful learning experiences. In 2050, technological advancement has also made teacher professional development in STEM education a whole easier. Remote access to laboratory experiments and makerspaces is possible and hence, teachers can work in teams regardless of their geographical location. Teachers are actively experimenting with different STEM pedagogies and assessment practices and hence STEM self-efficacy of both teachers and students is high. Instead of being mathematics, biology, physics or chemistry teachers, teachers are proudly adopting the identity of a teacher of STEM. Perhaps one research that would be of interest to you in 2050 is to compare the teaching and learning STEM education in 2023 and in 2050. What are some problems of integrated STEM teaching that still remain? How have assessment practices changed in 2050? What are the characteristics of questions that students ask in 2050? Are more students taking up courses of study in STEM at the university level? Have STEM literacy levels increased in 2050 compared to 2023? Have physical classroom infrastructures changed in 2050? Have we managed to slow down environmental degradation using knowledge in STEM? What sort of issues have arisen due to our switch to sustainable clean energy? Have scientists managed to innovate ways to treat plastic wastes in a less harmful manner? What lessons that we have learnt in STEM education in 2023 are still relevant in 2050? We hope that the work done by researchers and scholars in 2023 has been useful to help you move your knowledge forward. We would be thankful if is chapter could be taken as classical knowledge and reading as you look back at how integrated STEM learning experiences have evolved in last 30 years.

References

Anderson LW, Krathwohl DR, Airasian PW et al (eds) (2001) A taxonomy for learning, teaching, and assessing, Abridged. Addison Wesley Longman, New York

Barden LM (1995) Effective questioning & the ever-elusive higher-order question. Am Biol Teach 57(7):423–426

Bielik T, Yarden A (2016) Promoting the asking of research questions in a high-school biotechnology inquiry-oriented program. Int J STEM Educ 3(15):1–13. https://doi.org/10.1186/s40594-016-0048-x

Biggs J, Collis K (1982) Evaluating the quality of learning: the SOLO taxonomy. Academic Press, New York

Bloom BS (1956) Taxonomy of educational objectives. Handbook: the cognitive domain. David McKay, New York

Brand BR (2020) Integrating science and engineering practices: outcomes from a collaborative professional development. Int J STEM Educ 7(13):1–13. https://doi.org/10.1186/s40594-020-00210-x

Buchanan R (1992) Wicked problems in design thinking. Des Iss 8(2):5–21

Chin C (2007) Teacher questioning in science classrooms: approaches that stimulate productive thinking. J Res Sci Teach 44:815–843

Chin C, Osborne J (2008) Students' questions: a potential resource for teaching and learning science. Stud Sci Educ 44(1):1–39. https://doi.org/10.1080/03057260701828101

Chin C, Osborne J (2010) Supporting argumentation through students' questions: case studies in science classrooms. J Learn Sci 19(2):230–284. https://doi.org/10.1080/10508400903530036

Crismond DP, Adams RS (2012) The informed design teaching and learning matrix. J Eng Educ 101(4):738–797

Cuccio-Schirripa S, Steiner HE (2000) Enhancement and analysis of science question level for middle school students. J Res Sci Teach 37:210–224

Cuthbert AS (2021) Disciplinary knowledge and its role in the school curriculum. In: Cuthbert AS, Standish A (eds) What should schools teach?: Disciplines, subjects and the pursuit of truth, 2nd ed. UCL Press, pp 15–37. https://doi.org/10.2307/j.ctv14t475s.8
Dewey J (1938) Logic: the theory of inquiry. Henry Holt and Company Inc., New York
Eliasson N, Karlsson KG, Sorensen H (2017) The role of questions in the science classroom—How girls and boys respond to teachers' questions. Int J Sci Educ 39(4):433–452
Erdogan I, Ciftci A, Topcu MS (2017) Examination of the questions used in science lessons and argumentation levels of students. J Balt Sci Educ 16(6):980–993
Ford MJ (2015) Educational implications of choosing "practice" to describe science in the next generation science standards. Sci Educ 99(6):1041–1048
Herranen J, Aksela M (2019) Students-question-based inquiry in science education. Stud Sci Educ 55(1):1–36. https://doi.org/10.1080/03057267.2019.1658059
Honey M, Pearson G, Schweingruber H (eds) (2014) STEM integration in k-12 education: status, prospects, and an agenda for research. The National Academies Press
Huang X, Lederman NG, Cai C (2017) Improving Chinese junior high school students' ability to ask critical questions. J Res Sci Teach 54(8):963–987
King A, Rosenshine B (1993) Effects of guided cooperative questioning on children's knowledge construction. J Exp Educ 61:127–148
Kirschner PA, Sweller J, Clark RE (2006) Why minimal guidance during instruction does not work: an analysis of the failure of constructivist, discovery, problem-based, experiential, and inquiry-based teaching. Educ Psychol 41:75–86
Lai M, Law N (2013) Questioning and the quality of knowledge constructed in a CSCL context: a study on two grade-levels of students. Instr Sci Int J Learn Sci 41:597–620
Lederman NG, Abd-El-Khalick F, Bell RL, Schwartz RS (2002) Views of nature of science questionnaire: toward valid and meaningful assessment of leaners' conception of nature of science. J Res Sci Teach 39(6):497–521. https://doi.org/10.1002/tea.10034
Marbach-Ad G, Sokolove PG (2000) Good science begins with good questions. J Coll Sci Teach 30:192
Mortimer EF, Scott PH (2003) Meaning making in secondary science classrooms. Open University Press, Buckingham
National Research Council (2012) In: Press NA (ed) A framework for K-12 science education: practices, crosscutting concepts, and core ideas. Committee on a Conceptual Framework for New K-12 Science Education Standards. Board on Science Education, Division of Behavioral and Social Sciences and Education, Washington, DC
Osborne J, Reigh E (2020) What makes a good question? Towards an epistemic classification. In: Butler LP, Ronfard S, Carroveau KH (eds) The questioning child: insights from psychology and education. Cambridge University Press, Cambridge, pp 281–300
Shepardson DP, Pizzini EL (1991) Questioning levels of junior level high school science textbooks and their implications for learning textual information. Sci Educ 75:673–682
Tagnin L, Riordain MN (2021) Building science through questions in content and language integrated learning (CLIL) classrooms. Int J STEM Educ 8(34):1–14. https://doi.org/10.1186/s40594-021-00293-0
Tan A-L, Teo TW, Choy BH, Ong YS (2019) The S-T-E-M quartet. Innov Educ 1(3). https://doi.org/10.1186/s42862-019-0005-x
Teo TW, Tan AL, Ong YS, Choy BH (2021) Centricities of STEM curriculum frameworks: variations of the STEM quartet. STEM Educ 1(3):141
Watts M, Gould G, Alsop S (1997) Questions of understanding: categorising pupils questions in science. Sch Sci Rev 79(286):57–63
Wenning CJ (2009) Scientific epistemology: how scientists know what they know. J Phys Tchr Educ Online 5(2):3–14

Chapter 4
Epistemic Growth in Students' Understanding and Concern About Trust: A Practice-Oriented Approach to Learning Nature of Science

Jessica Shuk Ching Leung

4.1 Introduction

In this era of pandemics and infodemics—an overabundance of information and mis-information– has presented students with challenges in making informed judgment about knowledge claims in science on issues such as vaccines, lab-grown meat, microplastics, climate change, and many more. As scientists depend on highly specialized disciplinary knowledge to reach their conclusions, laypeople increasingly have to place their trust on judgments made by the scientific community as they have limited capacities to take up the role of specialists in evaluating scientific claims (Bromme and Goldman 2014). This predisposes laypeople, who are outsiders to science, to develop epistemic dependence on insiders of science (Hardwig 1985). Epistemic dependence entails a rejection of intellectual self-autonomy (Hardwig 1985; Zagzebski 2012), which means that one reckons his/her epistemic status as lower than a trusted person hence coming under the trusted person's authority for a particular view. Epistemic dependence requires trust while trust serves as a means to manage dependence (Wagenknecht 2015). Epistemic dependence is not passive; it requires one to be sceptical about assigning trust to knowledge claims. To become competent outsiders to science for making informed judgement about knowledge claims in science, students should therefore learn when and how to put trust in science (Fensham 2014).

When assigning trust to knowledge claims, outsiders to science should not feel the need to judge the claims directly, but to judge the grounds for their claims (Norris 1995). Rudolph and Horibe (2016) echoed this notion and described the former as "first-order" evaluations of scientific knowledge and the latter as "second-order"

J. S. C. Leung (✉)
Faculty of Education, The University of Hong Kong, Hong Kong, China
e-mail: leungscj@hku.hk

Y. S. Ong et al. (eds.), *A Diversity of Pathways through Science Education*,
https://doi.org/10.1007/978-981-97-2607-3_4

evaluations. The latter entails recognizing which individuals and research communities to rely on for relevant knowledge. These also often require an understanding of the peer review process, nature of expertise, conflicts of interest, and tentativeness that are embedded in the generation of scientific knowledge within the science community (Höttecke and Allchin 2020; Leung 2020a). Making use of these aspects of nature of science (NOS) was shown to be within the reach of college and university students too (e.g., Leung et al. 2017; Leung 2020b). Learning NOS involves understanding "how science works with the goal of interpreting the reliability of scientific claims in personal and public decision making" (Allchin 2011, p. 521). The way that NOS is characterized depends on what it takes to judge the reliability of these claims, and the list of NOS components is therefore "expansive, open-ended, and inclusive" (Allchin 2017, p. 20).

4.1.1 Learning NOS Through a Practice-Based Approach

Previous research on learners' grasp of NOS has mostly conceptualised NOS understanding as a form of knowledge (e.g., Schwartz et al. 2012) that can be accessed through learners' self-reported beliefs or views about NOS (Ford 2008). Yet, there is little evidence showing the mere recall or comprehension of NOS tenets is sufficient for their effective application in context (Allchin 2011). Recent findings have illustrated the difficulties that students encounter in translating their NOS understanding as a declarative form of knowledge into practice (Leung 2020a). As Ford (2008) points out, "given the driving concern with NOS understanding for citizenship abilities, it seems reasonable to view NOS understanding … as an active ability rather than passive declarative knowledge" (p. 172). He further argued that to develop such active ability, practice is essential. This argument clearly raises the need to distinguish between developing NOS understanding as a declarative form of knowledge and as a knowledge-embedded behaviour or practice. While these differences may seem subtle, they have significant implications for how NOS understanding should be developed in the classroom. An extensive body of literature has aimed at developing students' NOS understanding through an explicit reflective teaching approach (e.g., Khishfe and Abd-El-Khalick 2002; Lederman 2007). Yet these studies were limited to conceptualising NOS as a form of knowledge. As argued by Yacoubian (2015), students need to "be guided to (a) develop their NOS understandings and (b) apply them by making explicit links between their NOS understandings and the decisions that they would make on socioscientific issues" (p. 251) i.e., the knowledge-embedded practice approach to achieve NOS understanding. This argument was supported by empirical findings reporting a significant shift in the use of NOS understanding among university students through a practice-oriented approach to learning NOS (Leung 2020c). This approach to learning NOS was informed by the Apt-AIR framework of epistemic growth (Barzilai and Chinn 2018). Apt-AIR is derived from the Multifaceted Framework of Epistemic Thinking (Barzilai and Zohar 2014) and the AIR model (Chinn and Rinehart 2016) that identifies three epistemic components:

epistemic Aims and value (i.e., goals set by the actors and their perceived importance); epistemic Ideals (i.e., criteria that people use to evaluate whether epistemic goals have been met); and Reliable epistemic processes (i.e., procedures or strategies to achieve epistemic goals and produce epistemic products). Apt-AIR highlights that students' engagement with epistemic aims, ideals, and processes involves five interdependent aspects as shown below:

1. Be engaged in cognitive processes that reliably achieve epistemic aims in accordance with epistemic ideals.
2. Achieve epistemic aims adaptively in diverse situations.
3. Use epistemic metacognition to guide epistemic performance.
4. Experience motivational-affective dispositions to achieve valuable epistemic aims.
5. Achieve epistemic aims with other people.

Each of these aspects involves competent engagement with reliable epistemic processes (R) to achieve epistemic aims and values (A) in accordance with epistemic ideals (I). The epistemic aim of focus in this study was to assign informed trust to scientific claims by knowledgeable others (A) through evaluations that require using multiple aspects of NOS that are relevant to the issue at hand (I) and competent engagement with reliable epistemic processes (R). Based on the Covid-19 pandemic, recent work by the author goes into considerable detail that highlights four reliable epistemic processes for laypeople to ground their trust including (1) identifying whether recognition from peer reviewers has been obtained; (2) examining the credentials of those who claim expertise; (3) determining the level of expert consensus; (4) identifying possible sources of bias (Leung and Cheng 2021). These processes entail a good understanding of NOS and the use of such understanding.

According to the Apt-AIR framework, epistemic growth is not only about epistemic performance (e.g., the extent to which trust is well-grounded), but is also about understanding the performance (e.g., understanding what it takes to achieve apt performance and what makes a performance apt) and concern about epistemic performance (e.g., valuing of epistemic aims and the reliable processes for achieving these aims and inclination to pursue epistemic aims). Prior studies have reported significant shifts in student use of NOS after experiencing a practice-oriented approach to learning NOS (Leung 2020b, c). However, these studies tended to focus more on epistemic performance per se than students' understanding and concerns about their epistemic performance. While it is possible to achieve good epistemic performance without understanding about the performance, the latter is essential for effectively regulating one's epistemic performance. Concerns about epistemic performance is important for it affects one's inclination to pursue epistemic aims and the motivation in advancing epistemic performance. These metacognitive and motivational-affective aspects of epistemic growth go beyond a "cold" cognition dimension and would deepen our understanding about how a practice-oriented approach to learning NOS might shape students' epistemic growth. Contextualized within the Covid-19 pandemic, the study was guided by the following question: How did students reflect on changes in their understanding and concern about placing epistemic trust? What were the challenges to placing epistemic trust following a practice-oriented approach to learning NOS?

4.2 Method

This study utilized a qualitative case study approach with descriptive and interpretive emphases (Merriam 2016). This approach was deemed as useful to generate a rich, nuanced data set that offered insights to address the research questions in this study. Details of the study context, the practice-oriented approach coursework, as well as the data collection and analysis approaches are described as follows.

4.2.1 Study Context

73 undergraduate students enrolled in a general education course from a university in Hong Kong participated in this study. The course focusses on sense making of science-related social issues by enhancing students' understanding about science by enabling them to critically evaluate science-related social issues. The course had no science prerequisites and attracted students pursuing degrees in the arts, business administration, education, journalism, law, medicine, nursing, science, and the social sciences. Prior studies (e.g., Akgun and Kaya 2020; Liu and Tsai 2008) reported that university science majors did not necessarily possess better understandings about science than their non-science counterparts and there is no evidence that science majors in Hong Kong are exception.

The course comprises three key modules: (1) "Features of science", (2) "The making of science-related news in the media", and (3) "Critical evaluation of socio-scientific issues." The first two modules focus on the nature and limitations of science and science news and their generation processes, and the third module focuses on engaging with science-related social issues. "Trust in the time of corona" was a unit in the last module. The unit was guided by the Apt-AIR framework. Aspects of the Apt-AIR framework and Reliable epistemic processes (Table 4.1) are flagged in the lesson description in the following section. Covid-19 was chosen as a context to foster students' placing of informed epistemic trust for its timeliness, relevance and impact as well as its rapid scientific advancement and the widespread of its information in the media. These presented an urgent need for students to be able to place informed trust on knowledge claims (Aspect 4).

Table 4.1 Reliable epistemic processes for grounding trust (Duncan et al. 2018; Leung and Cheng 2021)

Reliable epistemic processes	Description
Peer review	To examine if publication platforms and dissemination channels include mechanisms for evaluation, vetting or filtering
Professional recognition	To identify who the experts are, including the level and relevance of their expertise
Expert consensus	To determine whether there is consensus in the scientific community; to identify areas of agreement and disagreement among experts
Conflict of interest	To identify sources of bias

4.2.2 The Covid-19 Lesson

The data reported in this study comes from the Covid-19 lesson described as follows. The lesson started with a class poll, followed by two case studies, namely "The association between 5G technology and Covid-19" and "A crisis of trust—the World Health Organisation (WHO)" (Leung and Cheng 2021), and ended with a hands-on activity to practice the placing of trust regarding knowledge claims about Covid-19.

4.2.2.1 Class Poll

Students were asked to indicate their response according to a 6-point Likert-scale ("Strongly Agree to Strongly Disagree") on the following questions: "Do you agree that 5G causes Covid-19? Do you agree that the WHO is trustworthy?" This was followed by small group discussions for students to make sense of the class poll results (Aspect 5) and set the stage for subsequent case studies.

4.2.2.2 The Association Between 5G Technology and Covid-19

The case study started by watching a video from the BBC titled "The people who think coronavirus is caused by 5G" (https://www.bbc.com/news/av/stories-53285610) to introduce students to rumors about the link between 5G technology and Covid-19 in the UK and its consequences such as arson attacks on cell phone towers and attacks on telecoms workers. In fact, similar claims about the health implications of telecommunications networks are not new: doctors experienced "radiophobia" as early as 1903, and there were widespread fears in the 1990s that radiation from 2G telecommunications technology could cause cancer. A viral video about popping corn with cell phones (https://www.youtube.com/watch?v=V94shlqPlSI) was used for illustration. Students were asked "Do you trust that we can pop corn with cell phones?" and why. A video showing CNN's interview with Abraham Glezerman,

CEO of the Cardo Systems, a manufacturer of Bluetooth headsets, was previewed. He admitted that the viral video had indeed been a marketing ploy. This reminded students to consider possible sources of bias or hidden interest when assigning trust (R, conflict of interest) (Aspect 1).

Students were then introduced to the news about the retracted paper titled "5G Technology and Induction of Coronavirus in Skin Cells" that was indexed in PubMed. This made the placing of trust problematic (i.e., whether to trust the rumors and the authors of the retracted paper, or other scientists). As the students were already introduced to the peer review process in the first module of this course, the question on whether peer-reviewed articles were worthy of trust was raised and discussed. Students were guided to examine the credentials of the authors and the journal (R, professional recognition and peer review) (Aspect 1). This case study concluded with the following questions in relation to epistemic processes for assigning informed trust (Aspect 3):

- Is hidden interest involved?
- Is there information about who did the study, where the study was done and how the results were made public?
- Do those who claim experts have the relevant expertise to make relevant claims?
- Has any unbiased source actually carried out a rigorous independent test of the claim and published the results, ideally in a reputable, peer-reviewed research journal?

4.2.2.3 A Crisis of Trust-WHO?

This case study started with drawing students' attention to a local news article titled "Coronavirus: World Health Organization faces credibility crisis among Hongkongers—survey" (https://hongkongfp.com/2020/02/25/coronavirus-world-health-organization-faces-credibility-crisis-among-hongkongers-survey/). It reported that 65% of respondents did not trust the WHO according to a telephone poll conducted among 576 Hong Kong residents (Aspect 4). The track record of WHO was also considered. During the Covid-19 outbreak up to March 2020, the WHO did not advise healthy people to wear masks unless they were infected or took care of the infected. Disappointedly, the precautionary principle—"in the case of serious or irreversible threats to the health of humans or the ecosystem, acknowledged scientific uncertainty should not be used as a reason to postpone preventive measures" (Martuzzi and Tickner 2004, p. 1)—that the organization has been abiding to was not reflected in this recommendation. Students were then guided to identify the possible sources of bias (R, conflict of interest) (Aspects 1 and 2) that might have caused the WHO to deviate from their own precautionary principle. Another track record of the WHO in the Covid-19 outbreak was about the comment by Dr. Mara Van Kerkhove (Head of the WHO emerging diseases and zoonosis unit) who said that asymptomatic transmission was 'very rare.' This was very different from a published study estimating that about 40% to 45% of people infected with Covid-19 would remain asymptomatic and would unknowingly spread the virus (Oran and

Topol 2020). Many health professionals raised concerns about this comment and Kerkhove later expressed regret over her earlier comment. In this scenario, it would be useful to identify whether recognition from peer reviewers had been obtained (R, peer review) and to determine the level of expert consensus (R, expert consensus) (Aspect 1). This case study scaffolded and extended students' understanding of the epistemic processes (Aspect 2), including: Is there a proven track record by the one who claimed expertise? Is there information about what other scientists think? (Aspect 3).

4.2.2.4 Let's Practice!

Lastly, the students worked in small groups on a news reading task. They were presented with a news article titled "Coronavirus: India must cut pollution to avoid Covid disaster" (https://www.bbc.com/news/world-asia-india-54596245). They were asked to what extent they agreed that the severity of Covid-19 infection was associated with exposure to air pollutants and to identify criteria that they had considered in reaching their decision by reflecting upon their decision-making process (Aspects 2, 3 and 5).

4.2.3 Data Collection and Analysis

Upon completing the lesson, students were asked to reflect on their learning experience by writing a reflective journal. This reflective writing task served as a data source to answer the research questions. The coding and analysis process involved a few rounds of iterations and refinement between the author and a fellow researcher. The data were analysed using codes that were inductively inferred over a few rounds of iterations. First, the two coders examined the data independently before meeting to identify students' reflections on their understanding and concern about placing epistemic trust and challenges they faced. Then, a list of codes was generated to characterize the reflections. Next, the coders returned to the data for another round of independent analysis, during which the data were re-examined, and the list of codes was refined. The following three themes and their corresponding number of codes emerged from the data: *Growth in understanding about placing epistemic trust* (three codes) *Concern about placing epistemic trust* (two codes) and *Challenges to placing epistemic trust*(two codes). Details of the codes are elaborated in the following section. Interrater reliability by the two coders was greater than 0.85, as measured by Cohen's kappa coefficient. Discrepancies were discussed until consensus was reached.

4.3 Results

In summary, the students' reflective journals indicated the following aspects about their epistemic trust following their experience of the Covid-19 lesson. Firstly, students demonstrated *growth in their understanding about placing epistemic trust* through (i) better understanding of the need and reasons of assigning informed trust to experts, (ii) consideration of a broader range of epistemic processes, and (iii) increased consciousness about the epistemic processes of placing trust. Secondly, students demonstrated *greater concern about placing epistemic trust* as indicated by (i) increased concern and inclination in placing epistemic trust and (ii) increased self-efficacy in achieving epistemic aim and its appreciation. However, students also reflected *challenges to placing epistemic trust*, as some students indicated a lack of self-efficacy and intention to pursue epistemic aims while others expressed the lack of a perfect solution on who/what to trust.

4.3.1 Growth in Understanding About Epistemic Trust

4.3.1.1 A Better Understanding of the Need and Reasons of Assigning Informed Trust to Experts

From their reflections, some participants expressed blindly assenting to knowledgeable others prior to the unit:

> In the past, I tended to *blindly believe* in every news and study I read. I thought all the writers of the research studies were professional and skilled... Now, I understood that there are many factors that lie behind the publication of studies. For instance, there may be *hidden intentions* masked behind the study or the researcher may not be experienced in the topic of his/her research. Hence, I will *think twice* and examine the *background of the author* before believing the information. (Andrew, emphasis added)

Andrew (all names used here are pseudonyms) reflected on his shift from "blindly believing in every news and study" to "think[ing] twice." Such change was brought by his understanding of the "many factors" involved in the publication of studies, e.g., "hidden intentions" and the track record of authors ("the researcher may not be experienced in the topic"). This suggested that understanding the social aspects of science, e.g., the human factors involved in the doing of science and the reporting of its findings, could be attributed to this change. This was also noticeable in Amy's excerpt:

> ... I used to *trust every news or information* that I read because they were always written in a professional tone by some "experts". But after today's lesson, I realized that we *need to be aware of which kind of experts* they are because their expertise may not be so relevant to the topic. (Amy, emphasis added)

Amy also reported a similar shift from "trust[ing] every news or information" to realizing "the need to be aware of which kind of experts they are" for the implication of authors' relevance of expertise. The above suggested that understanding the social aspects of science, e.g., hidden interests, track record and relevance of expertise, assisted in the awareness of the need for laypeople to assign informed trust to experts. The following excerpts suggested that such understanding may also lead to the growth of understanding *why* it was necessary to assign informed trust to experts:

> To gain reliable knowledge on a certain topic, asking experts in that field seems like the best choice. However, it *does not mean that they are always right*. Occasionally, some highly respected authority is *dishonest* and claim to know everything or may make *mistakes*. (Benjamin, emphasis added)
>
> WHO, though authoritative as it appears, is still comprised of *humans*. *Mistakes* can be made, notwithstanding the amount of expertise it has. Thus... it is important to acknowledge the *possible errors*, and not treat it as a "bible" at our use. We should always ponder its motive, hidden interest, evidence etc. before placing trust. (Beatrice, emphasis added)

Benjamin and Beatrice not only acknowledged the need to place informed trust (e.g., "it does not mean that they are always right" and "not treat it as a bible"), but also the reasons why, i.e., the human factors at play ranging from dishonesty, mistake, motive to hidden interest.

4.3.1.2 A Consideration of a Broader Range of Epistemic Processes

While some participants reported a shift from blindly assenting to experts to identifying the need and reasons for assigning trust to experts, others considered a broader range of epistemic processes in their placing of trust:

> I am delighted that the lecture echoes my concerns about how to prevent myself from questioning everything and finally trusting nothing in the post-trust era. I *used to end my fact check* when I found the assertions have scientific endorsement, but now *I'll continue to ask questions* about the journal's credibility and the scientist's expertise. (Chloe, emphasis added)
>
> Usually when I look at a piece of information, I would not consider the factor of *hidden interest* or examine thoroughly the proposed explanation before placing trust on it. However, these omitted steps might cause a big deal in determining whether the news is trustworthy or not, this reminds me of asking critical question before placing trust. (Charles, emphasis added)

Chloe reported an extension in her "fact check" from "scientific endorsement" to "journal credibility and scientist's expertise" while Charles reported his appreciation of hidden interest that he would not have considered previously. Both suggested considering a broader range of epistemic processes after the lesson.

4.3.1.3 Increased Consciousness About Epistemic Processes of Placing Trust

Participants also reported that they became conscious about their practice of assigning trust to knowledgeable others as exemplified in the following excerpts:

> I usually trust someone or something based on my personal perception *without setting up any criteria* such as those critical questions raised in the lesson. Those questions help us consider more about trust in a more *rational* way. For example, the case of popping corn by using mobile phone seems real. However, it is a joke and there is *hidden interest* associated with promoting the Bluetooth headsets. Hence, they [the case studies] show the *importance* of setting some criteria and raising critical questions before placing trust. (Emma, emphasis added)

> Before this lecture, I was *not aware* of how I assigned trust to different sources. I might tend to rely on official reports issued by the Legislative Council, or journals written by "tycoon" scholars on a specialized field. After this lecture, it becomes clearer to me that I do have a hidden list of criteria for assigning trust. I *value* the reputation of the contributors, their officiality and professionalism, and the journal quality. (Ethan, emphasis added)

Both Emma and Ethan expressed that they became more conscious about their strategies in assigning trust to knowledgeable others. At post-lesson, not only did they express an understanding of reliable epistemic processes (e.g., hidden interest and expertise), but also an appreciation of these processes, e.g., "importance of setting some criteria and raising critical questions" (Emma) and "I *value* the reputation of the contributors…" (Ethan). This seemed to suggest an association between understanding and being concerned about these processes (to be further discussed in the next section).

4.3.2 Growth in Concern About Placing Epistemic Trust

4.3.2.1 Concern and Inclination in Placing Epistemic Trust

Participants reflected on their increased concern and inclination in placing trust:

> At the beginning of the lesson, I thought placing trust is a *minor issue*, as it does not affect much about the nature and development of the event. It was until that moment, I found out placing trust on certain issues can bring *greater impacts than I could have imagined*. At times, it is even our responsibility to believe in the right side and I *can never overlook the importance of it* [placing trust]… To make sound judgments, it is *crucial* to keep asking questions. From the hidden interests and proposed mechanisms, to the reputation and the arguments of the information, these all can be *vital* factors for us to make decisions. In the end, placing trust is *much harder* than I have imagined and it is a *lifetime process to learn about that*. (Nicholas, emphasis added)

Nicholas reflected on his shift from viewing placing trust as "a minor issue" to "never overlook[ing] its importance" for the "greater impacts [of placing trust] than [he] could have imagined." This suggested that students rarely reflected on the importance of placing trust and consequences of misplacing trust. Engaging students in authentic cases (e.g., the 5G case) of misplacing trust and its adverse consequences could be a starting point for students to reflect on the importance of placing informed trust. Nicholas did not only care about placing trust more, but also its reliable epistemic processes. For instance, he perceived it "crucial" to consider "hidden interests," "proposed mechanisms," "reputation and the arguments of the information." Notably, despite perceiving placing trust as "much harder" than imagined, Nicholas expressed his willingness and intention to overcome such challenge by viewing it as a "lifetime process to learn about that."

Echoing Nicholas, both Naomi and Nathan expressed that the impact of misplacing trust was more serious than they could have imagined:

> ... what *shocked* me most was that there are people, who are very convinced by it [the 5G conspiracy theory]... However, when I reflect upon it, it is entirely understandable. As a social media user, I have also reposted posts from accounts that I believe are credible without thoroughly researching on. It is easy for us to spread news by just clicking a button, but it is in fact *very dangerous* to do so without confirming the source as people who are under fear or confusion would not be able to think rationally and this may lead to *greater negative impact* than we have imagined. (Naomi, emphasis added)
>
> ... When I was reading the news about the distrust of 5G in some countries, it was *like a joke* to me, but as I reconsider the local situation, this laziness to review propaganda's trustworthiness is *hitting harder than I have imagined*. For example, the scandal of portable air purifier which claimed to equip the wearer with a virus-free barrier. Some citizens of Hong Kong did believe that and purchased that, even my mother has thought of buying it. The hidden interest, profit made, is apparent. Yet it is interesting to consider a local case of how easily we can place our blind trust on some seemingly scientific products. The 5G incident is nowhere to be seen in Hong Kong, but it exists in other forms. It is unmanageable for us to debunk every myth, but it is possible for me to work with others to debunk myths. Ever since my mum's wish for a portable air purifier arises, I have been teaching her to look into details of the propaganda and to discuss with people *outside her echo-chamber* about the product... (Nathan, emphasis added)

The way the public responded to the 5G conspiracy theory came as a surprise to both students ("shocked"; "like a joke"). Drawing on her practice of reposting posts, Naomi concluded that "it is entirely understandable." She realized how easy it was to spread news and how dangerous it was for its negative impacts. Similarly, Nathan related the distrust of 5G to the mistrust of air purifier in the local context. He commented that the impact was "hitting harder than [he had] imagined". Both participants related the 5G conspiracy theory to their everyday life, expressing their concern about misplacing trust for its negative impacts.

Nathan, while acknowledging the difficulty of "debunk[ing] every myth," viewed it as within his reach to work with others to overcome the difficulty, e.g., through "look[ing] into details of the propaganda" and "discuss[ing] with people outside [their] echo-chamber." This suggested his practicing of placing informed trust. In summary, the above reflections not only highlighted participants' increased concern about placing trust and the associated epistemic processes, but also their intention to further learn about it and their practicing of placing trust in everyday life.

4.3.2.2 Self-efficacy in Achieving Epistemic Aim and Its Appreciation

It is essential for students not only to recognize the need, but also to feel empowered in placing trust. Patrick explained how laypeople could assign informed trust in the midst of contradictory claims in real life:

> As an *ordinary person* without profound scientific knowledge, I have to rely on *professionals* for information. However, we are often *confused* because there are too many versions of explanations from "professionals," sometimes they are contradictory! .... The trouble comes: Which version should we trust? There are *ways for us to make a sensible judgement*, say, is there sufficient evidence to support the claim? Is the claim widely agreed by other scientists? Or just basically, does it make sense? (Patrick, emphasis added)

As "an ordinary person," Patrick's reflection suggested his self-efficacy in placing trust (e.g., "there are ways for us to make a sensible judgment") despite the contradictory claims among "professionals." Patrick not only suggested processes pertaining to a first-order evaluation ("is there sufficient evidence?"), but also a second-order evaluation ("is the claim widely agreed by other scientists?") in his assigning of trust. While Patrick demonstrated his self-efficacy, Phoebe showed her valuing of such capability:

> ... because until now we really do not understand Covid-19 very well, it would be very hard to judge these conclusions. In this case, we, as ordinary people, *really need to have some capabilities* to judge the quality of different journals, the qualification of writers, and even acquire some basic science knowledge to avoid following others' ideas blindly. (Phoebe, emphasis added).

Despite the challenge of judging knowledge claims about Covid-19 ("very hard to judge these conclusions"), Phoebe suggested epistemic processes for ordinary people to evaluate these claims, and also expressed her valuing of these "capabilities."

4.3.3 Challenges

Despite the epistemic growth reported above, learning to place trust was not without challenges. For instance, while some students demonstrated self-efficacy in placing trust, others did not:

> I feel *more uncertain* about who to trust. While impact factor may be useful, the work of famous scientists is more likely to be cited, which is unfair to the less well-known scientists. People also tended to cite paper with higher citation, therefore impact factor may not be that credible. (Ruby, emphasis added)

Unlike those who appreciated epistemic processes, such as considering the quality of journals (e.g., Phoebe), Ruby had reservations over the use of impact factors to reflect journal quality even when she viewed it as useful. The lesson did not intend to suggest impact factor as a silver bullet to assigning trust. It was possible that Ruby's uncertainty about who to trust was attributed to her over-reliance on impact factor as a single criterion in placing trust. Should Ruby recognize the epistemic ideal of considering multiple criteria for placing trust, this would likely ease her uncertainty about who to trust.

Like Nicholas and Phoebe, Ryan also acknowledged the importance and challenge of placing trust. However, he lacked the self-efficacy and intention in pursuing this aim:

> Although placing informed trust is crucial, it is hard to maintain in daily life. We are not able to judge and 'fact-check' everything to draw the correct conclusion at all. *All we can do is to choose something to believe.* (Ryan, emphasis added)

His reflection vividly portrayed what people in the infodemic era were facing—being bombarded with loads of information beyond their capacity "to judge and 'fact-check' everything." While it is hardly possible for individuals to assign informed trust on every single knowledge claim they encounter, being able to assign informed trust would still allow them to assess the degree of trust to place on those knowledge claims of importance and interest to them. Sadly, Ryan made the pessimistic remark, "all we can do is to choose something to believe," as if placing informed trust was out of the equation. In other words, despite valuing the placing of informed trust, he did not seem inclined to pursue it in his daily life. Similarly, Renee similalry made a pessimistic remark in her reflection:

> There is no "best" solution to any crisis in the world... Even for those reputable scientists, they are also human being and all humans may make mistakes. In my opinion, even though this is a *pessimistic* view, there is nothing that we can 100% trust in. Everything can change, but it is also the charm behind the world. What we need to do is to be responsible for what we have decided. (Renee, emphasis added)

Despite acknowledging the lack of "best" solution, the possibility for reputable scientists to make mistakes and the uncertainties associated with placing trust, Renee interestingly appreciated all these negative aspects by describing them as "the charm behind the world" and expressed the intention of bearing the responsibility of placing trust.

4.4 Discussion

Unlike prior studies on learning NOS that largely focused on students' performance per se, this study adds to the existing literature by looking into students' understanding and concern about their epistemic performance. This approach offers a metacognitive and motivational-affective lens to our current understanding about how a practice-oriented approach to learning NOS shapes epistemic growth. First, corroborating with empirical findings on the effect of a practice-oriented approach to NOS (Leung 2020c), participants in this study reported their consideration of a broader range of epistemic processes in their placing of trust. In addition, participants became more aware of the need of and underlying reasons for placing well-grounded trust. This finding from student perspective is a timely response to the call for a shift from a deficit model that assumes the public to be ignorant and passive beneficiaries of science to the model of public engagement with science, which assumes that citizens can play an active role in science through public participation (e.g., Irwin 2014; Stocklmayer and Bryant 2012; Sturgis 2014). It is not only about engaging students in shaping and evaluating knowledge within a community but is also about taking ownership and control of their epistemic agency. Second, it is evident from the data that understanding the social aspects of science may attribute to the increased awareness of the need and reasons for laypeople to place informed trust on experts. Third, participants reported increased consciousness about reliable epistemic processes of placing informed trust. This is important for the regulation of their use of appropriate processes for achieving apt epistemic performance (Barzilai and Chinn 2018).

Participants expressed a shift from taking placement of trust lightly to acknowledging its importance. Findings suggested that engaging students in discussion and reflection on the adverse impacts of misplacing trust might account for the shift as in the case of the 5G coronavirus conspiracy theory. Given that beliefs and thoughts are predispositions to actions (Ajzen and Fishbein 1973), the shift in acknowledging the importance of placing informed trust might lead to participants' inclination to pursue this epistemic aim. This was evidenced in Nicholas's inclination to further learning of placing trust and Nathan's practicing of trust-placing in everyday life. Furthermore, participants expressed their appreciation for reliable epistemic processes, e.g., considering the hidden interests and relevance of expertise, through which their epistemic aim of placing informed trust would be achieved. This suggested an association between participants' caring about epistemic aims and the reliable epistemic processes that enable them to achieve their epistemic aims.

Prior findings reported the development of self-efficacy through experiencing successful achievement of epistemic aims (Leung 2020b), but highlighted concern about the possibility that learners might become over-confident and over-simplify the complex process. Findings of this study might allay this fear, as evidenced in participants' reflection about the placing of trust as a challenging task, e.g., "harder than imagined" (Nicholas), "very hard to judge" (Phoebe). Apart from increased self-efficacy, this study also reported participants' appreciation for the capability of placing informed trust (e.g., Phoebe), which in turn, could act as a driving force to motivate students' learning of placing informed trust.

Participants expressing increased understanding about reliable epistemic processes often showing appreciation for these processes as well (e.g., Emma and Ethan). This suggested an association between participants' understanding and caring about reliable epistemic processes, lending empirical support to the interdependence between aspects in the apt-AIR framework (i.e., Aspects 3 and 4). On the one hand, it is possible that it requires an understanding about reliable epistemic processes for students to recognize the value of these processes; on the other hand, caring about epistemic aims may motivate students to learn about these reliable processes to achieve their epistemic aims. Future research for a more nuanced understanding to unpack the relationship among various aspects in the apt-AIR framework would advance our understanding about epistemic growth.

It is interesting to note that participants acknowledging the importance of and challenges associated with placing trust might view these challenges in different ways. For instance, Nicholas chose to take up the challenge and pursue the learning of placing trust as a lifetime process while Ryan viewed this challenge negatively, which might be, at least in part, due to his ambitious goal of judging and fact-checking *everything* he came across. Interestingly, Renee adopted a mixed view about the challenge. While she viewed nothing worthy of our complete trust, she perceived this as "the charm behind the world" and was upfront about taking full responsibility for the decisions she made. Placing epistemic trust is never straightforward and has proved to be a challenge to outsiders to science. Findings from this study captured the complexity in students' perception to the challenge of placing trust. The way students responded to this challenge was crucial to their engagement with placing epistemic trust in authentic contexts, e.g., whether students would carefully consider reliable epistemic processes or whether they would simply place their trust out of intuition. What accounts for the difference in participants' perception to the challenge? How to regulate students' pessimism to the challenge and synergize it for placing informed trust? Future research that explores student responses to the challenge of placing trust and their associations with epistemic performance would inform teaching designs.

4.5 Conclusion

4.5.1 A Practice Turn to Learning NOS

The covid pandemic is not the first nor will it be the last human crisis. Climate change, supply of fresh water, energy crisis, biodiversity, conservation, sustainable development, public health and more will continue to be pressing issues to mankind. Coupled with the spread of misinformation and disinformation (i.e., misinformation with intention to mislead) in this era of infodemic, this presents an unprecedented need for citizens to be able to assign informed trust on claims about these issues. It is more realistic for citizens, who are outsiders to science, to assign trust based on second-order evaluations of information than first-order evaluations. This necessitates learning NOS, which should not be conceived as a form of knowledge alone, but a form of knowledge-embedded practice. While an explicit reflective teaching approach has shown promise in developing students' NOS understanding as a form of knowledge (e.g., Khishfe and Abd-El-Khalick 2002; Lederman 2007), such an approach may not be equally effective in preparing students for effective application of NOS understanding in context. This calls for a practice turn to learning NOS that has been drawing and will continue to draw increasing attention (e.g., Leung 2020a, b, c; Yacoubian 2015).

4.5.2 Beyond a "Cold" Cognition Approach to Learning NOS

Most of the foregoing studies on learning NOS adopted a cold cognition approach without considering such factors as values, motivation, and self-efficacy. For those who do not acknowledge the need and value of placing informed trust, those who are not up to the challenge of placing trust and those who lack self-efficacy in placing trust, it is unlikely that they will pursue this epistemic aim in their everyday life.

Developing students' NOS understanding as a form of practice should go beyond a "cold" cognition approach. A warm trend in learning NOS that honours a practice turn will better prepare future generations most of whom will be outsiders to science. In this way, they will be able to place well-grounded trust for informed judgment about science-related claims in this and future eras of pandemics and infodemics.

4.6 Note to Our Future Colleagues

In today's infodemic era, the public is adversely influenced by the spread of scientific misinformation and disinformation in the media. Covid-19 is not the first nor will it be the last human crisis. The infodemic has continued to prevail as evidenced in the continued widespread of misinformation and disinformation about challenges surrounding public health, food, energy, natural resources, the environment, sustainable development, and more. Furthermore, the prevalence of Generative AI (GAI) has fueled the infodemic as GAI is capable of producing well-articulated, expert-like information that makes it challenging for the layperson to distinguish between facts and hallucinated information.

Around the time of the Covid-19 pandemic, there was much hype about the need to get our young students "future-ready". Decades later, there is growing consensus among education researchers and practitioners that learning and teaching NOS as a form of practice is a promising approach for preparing students to cope with challenges in this infodemic era. This approach necessitates a warm trend in learning NOS that includes and is not limited to students' valuing of and their self-efficacy in using NOS. Following the vast research on teaching and learning NOS as a form of knowledge in the past two decades or so, literature on the teaching and learning of NOS as a form of practice is predicted to gain momentum in the coming decades to better prepare students for the challenges to mankind.

Parallel to this warm trend in NOS, the role of emotions that plays into the placement of trust on knowledge claims around challenges to mankind warrants our attention. In evaluating these knowledge claims, people may experience a mix of fear, helplessness, frustration, panic, guilt, shame, hopefulness, empathy, and more. How do these emotions impact students' epistemic performance in placing trust? Specifically, how do these emotions impact multiple aspects of epistemic performance in placing trust, including valuing of and self-efficacy in using NOS? There has been increasing attention to the complexity of emotions in students' engagement with knowledge claims around these challenges to mankind. This growing body of literature has started to shed light on how emotions may influence and be influenced by the cognitive, metacognitive, motivational and social aspects of placing trust. Further research on how to support students in regulating their emotions and harnessing them for apt epistemic performance in placing informed epistemic trust is needed. Through continued research efforts in these areas, the students of today will become responsible citizens of tomorrow competent in making informed decisions amidst any infodemic.

Acknowledgements This study is based upon work supported by the General Research Fund of the Research Grants Council under Project Number 17609121.

References

Ajzen I, Fishbein M (1973) Attitudinal and normative variables as predictors of specific behaviors. J Pers Soc Psychol 27(1):41–57. https://doi.org/10.1037/h0034440

Akgun S, Kaya E (2020) How do university students perceive the nature of science? Sci Educ 29(2):299–330. https://doi.org/10.1007/s11191-020-00105-x

Allchin D (2011) Evaluating knowledge of the nature of (whole) science. Sci Educ 95(3):918–942. https://doi.org/10.1002/sce.20432

Allchin D (2017) Beyond the consensus view: whole science. Can J Sci Math Technol Educ 17(1):18–26. https://doi.org/10.1080/14926156.2016.1271921

Barzilai S, Zohar A (2014) Reconsidering personal epistemology as metacognition: a multi-faceted approach to the analysis of epistemic thinking. Educ Psychol 49(1):13–35. https://doi.org/10.1080/00461520.2013.863265

Barzilai S, Chinn CA (2018) On the goals of epistemic education: promoting apt epistemic performance. J Learn Sci 27(3):353–389. https://doi.org/10.1080/10508406.2017.1392968

Bromme R, Goldman SR (2014) The public's bounded understanding of science. Educ Psychol 49(2):59–69. https://doi.org/10.1080/00461520.2014.921572

Chinn CA, Rinehart RW (2016) Epistemic cognition and philosophy: developing a new framework for epistemic cognition. In: Greene JA, Sandoval WA, Bråten I (eds) Handbook of epistemic cognition. Routledge, New York, NY, pp 460–478

Duncan RG, Chinn CA, Barzilai S (2018) Grasp of evidence: problematizing and expanding the next generation science standards' conceptualization of evidence. J Res Sci Teach 55(7):907–937

Fensham PJ (2014) Scepticism and trust: two counterpoint essentials in science education for complex socio-scientific issues. Cult Sci Edu 9(3):649–661. https://doi.org/10.1007/s11422-013-9560-1

Ford M (2008) 'Grasp of practice' as a reasoning resource for inquiry and nature of science understanding. Sci Educ 17(2–3):147–177. https://doi.org/10.1007/s11191-006-9045-7

Hardwig J (1985) Epistemic dependence. J Philos 82(7):335–349. https://doi.org/10.2307/2026523

Höttecke D, Allchin D (2020) Reconceptualizing nature-of-science education in the age of social media. Sci Educ 104(4):641–666. https://doi.org/10.1002/sce.21575

Irwin A (2014) From deficit to democracy (re-visited). Public Underst Sci 23(1):71–76. https://doi.org/10.1177/0963662513510646

Khishfe R, Abd-El-Khalick F (2002) Influence of explicit and reflective versus implicit inquiry-oriented instruction on sixth graders' views of nature of science. J Res Sci Teach 39(7):551–578. https://doi.org/10.1002/tea.10036

Lederman NG (2007) Nature of science: past, present, and future. In: Abell SK, Lederman NG (eds) Handbook of research on science education. Lawrence Erlbaum, pp 831–879

Leung JSC (2020a) Students' adherences to epistemic understanding in evaluating scientific claims. Sci Educ 104(2):164–192. https://doi.org/10.1002/sce.21563

Leung JSC (2020b) Promoting students' use of epistemic understanding in the evaluation of socio-scientific issues through a practice-based approach. Instr Sci 48(5):591–622. https://doi.org/10.1007/s11251-020-09522-5

Leung JSC (2020c) A practice-based approach to learning nature of science through socioscientific issues. Res Sci Educ. https://doi.org/10.1007/s11165-020-09942-w

Leung JSC, Wong ASL, Yung BHW (2017) Evaluation of science in the media by non-science majors. Int J Sci Educ Part B 7(3):219–236. https://doi.org/10.1080/21548455.2016.1206983

Leung JSC, Cheng MMW (2021) Trust in the time of corona: epistemic practice beyond hard evidence. Cult Stud Sci Educ 1–10. https://doi.org/10.1007/s11422-021-10045-9

Liu SY, Tsai CC (2008) Differences in the scientific epistemological views of undergraduate students. Int J Sci Educ 30(8):1055–1073. https://doi.org/10.1080/09500690701338901

Martuzzi M, Tickner JA (eds) (2004) The precautionary principle: protecting public health, the environment and the future of our children. World Health Organization, Copenhagen, Denmark

Merriam SB (2016) Qualitative research: a guide to design and implementation. Jossey-Bass

Norris SP (1995) Learning to live with scientific expertise: towards a theory of intellectual communalism for guiding science teaching. Sci Educ 79:201–217. https://doi.org/10.1002/sce.3730790206

Oran DP, Topol EJ (2020) Prevalence of asymptomatic SARS-CoV-2 infection: a narrative review. Ann Intern Med. https://doi.org/10.7326/M20-3012

Rudolph JL, Horibe S (2016) What do we mean by science education for civic engagement? J Res Sci Teach 53(6):805–820. https://doi.org/10.1002/tea.21303

Schwartz RS, Lederman NG, Abd-el-Khalick F (2012) A series of misrepresentations: a response to Allchin's whole approach to assessing nature of science understandings. Sci Educ 96(4):685–692. https://doi.org/10.1002/sce.21013

Stocklmayer SM, Bryant C (2012) Science and the public: what should people know? Int J Sci Educ Part B 2(1):81–101. https://doi.org/10.1080/09500693.2010.543186

Sturgis P (2014) On the limits of public engagement for the governance of emerging technologies. Public Underst Sci 23(1):38–42. https://doi.org/10.1177/0963662512468657

Wagenknecht S (2015) Facing the incompleteness of epistemic trust: managing dependence in scientific practice. Soc Epistemol 29(2):160–184. https://doi.org/10.1080/02691728.2013.794872

Yacoubian HA (2015) A framework for guiding future citizens to think critically about nature of science and socioscientific issues. Can J Sci Math Technol Educ 15(3):248–260. https://doi.org/10.1080/14926156.2015.1051671

Zagzebski L (2012) Epistemic authority: a theory of trust, authority, and autonomy in belief. Oxford University Press

Part II
Developing Science Teaching and Assessment

Chapter 5
Commentary for Part II: Two Fundamental Pillars in Science Education

Yew-Jin Lee

In this edited collection, Part II pulls together four substantive chapters that variously examine aspects of science teaching and assessment. Given this diversity of subtopics, it seems not unreasonable to enlist a few theoretical lenses to detect if there are any interesting trends or patterns to make sense of these chapters. Table 5.1 below is my attempt although I do acknowledge that another science educator might easily arrive at a different set of analytic ideas held together by a compelling narrative. In addition, readers might realize that my choice of concepts do not typically appear side-by-side in the scholarly literature as what has been done here. There is a reason for this: I recall the provocation by Giles Deleuze who argued that novelty or insight comes when we place disparate things/thoughts together, when we continue to add/edit/revise things that appear settled or closed to further interpretation (Deleuze and Guattari 1988). Hence, there are no closures because this interrupting process opens up new possibilities, visions, and relations that enrich life. For he and his collaborator Felix Guattari, the budding analyst should favor the conjunction AND over the verb IS (et[and] and *est*[is] are pronounced identically in French) because

> the AND gives relations another direction, and puts to flight terms and sets, the former and the latter on the line of flight which it actively creates. Thinking *with* AND, instead of thinking IS, instead of thinking *for* IS: empiricism has never had another secret. Try it, it is quite extraordinary thought, and yet it is life. (Deleuze and Parnet 2002, p. 43)

Based on Table 5.1, the second column on the left attempts to classify the chapters according to the chosen research design considerations, namely, by their major research focus/concern (viz. curriculum and/or pedagogy and/or assessment). We thus see that Chap. 6 made a dual contribution towards pedagogy as well as assessment when the three secondary science teachers involved in this project

Y.-J. Lee (✉)
National Institute of Education, Nanyang Technological University, Singapore, Singapore
e-mail: yewjin.lee@nie.edu.sg

Y. S. Ong et al. (eds.), *A Diversity of Pathways through Science Education*,
https://doi.org/10.1007/978-981-97-2607-3_5

Table 5.1 Some trends and patterns in the four substantive chapters in Part II across research design considerations and two theoretical lenses

Chapter	Research focus or concern	Main type of knowledge contribution to teaching	Major focus of institutional values
6	Pedagogy and Assessment	Knowledge for practice, Knowledge of practice	Academic intelligence
7	Assessment	Knowledge for practice	Academic intelligence
8	Assessment	Knowledge of practice	Academic intelligence
9	Curriculum and pedagogy	Knowledge for practice	Academic intelligence, Development of modern individuals as a collective good

utilized the RASA framework to improve their instruction and give targeted feedback on students' writing. Chapters 7 and 8, on the other hand, were strongly assessment-focused; the former sought to refine a framework for the intellectual or cognitive demands of structured items from the Singapore-Cambridge General Certificate of Education (GCE) O-Level examinations. The latter was an empirical examination of the reasoning/choices behind as well as the enactment of formative assessment practices in local primary science classrooms. Of course, knowledge derived from any single study in this category (and others) may suggest fresh insights for either curriculum or pedagogy, but their primary contributions were still located within the assessment domain. Likewise, Chap. 9 had a similar focus on curriculum and pedagogy when the authors took pains to describe the conceptual foundations behind their Integrated Science, Technology, Engineering, and Mathematics—Problem-Based Learning—Education for Sustainable Development (I-STEM-PBL-ESD) framework.

Another set of lenses to make sense of Part II is to consider how each chapter might serve the functions of increasing teacher knowledge, specifically *knowledge for, in,* and *of practice* (Cochran-Smith and Lytle 1999). Knowledge for practice denotes the generic principles, concepts, and theoretical commitments that comprise abstract knowledge surrounding teaching, pedagogical strategies, classroom management, and so forth. These understandings are most often formalized into forms for novices to learn during teacher training for subsequent application in the future. Knowledge in practice refers to the craft knowledge that teachers acquire chiefly through their years of work experience on the job as well as through countless interactions with pupils and other stakeholders. It thrives when practitioners critically reflect upon and deliberate on these diverse experiences although passing on such practical knowledge to others is not as straightforward as that of knowledge for practice. And when teachers (together with other stakeholders) conduct systematic investigations on teaching, a knowledge of practice develops. The latter merges practice and insights from the other two forms of teacher knowing because teachers now make "their classrooms and schools sites for inquiry, connecting their work in schools to larger issues, and taking a critical perspective on the theory and research of others" (Cochran-Smith and Lytle 1999, p. 273). If this third aspect of teacher knowledge is foregrounded, teachers are hence both consumers as well as producers of knowledge that generates useful insights for their own classrooms as much as teachers learn and contribute valuable, transformative information back to the (research, school, local) community (Darling-Hammond and Bransford 2005).

From Table 5.1, we thus observe that all chapters apart from Chap. 8 contribute towards knowledge for practice because they offer ample guidance and recommendations concerning what to teach and/or assess. Although they have drawn on varying extents of practical knowledge derived from classroom/course data, they are more rigorously characterized by the conceptual outcomes within their proposals. Knowledge of practice was exclusively found in Chap. 8 where the researchers investigated decision-making and the common practices related to assessment as part of a large baseline study of teaching in Singapore. Chapter 6 too contained elements of knowledge of practice for it was a conscious and collaborative effort by the authors to work with teachers through a professional development project. It is, of course, impossible to say which chapter created greater impact with respect to growing teacher knowledge; all the chapters have added much value towards the teaching profession in their own way.

The final lens that I wish to laminate over the four chapters comes from neo-institutionism. Here it is argued that contemporary educational systems share a small number of key features because of the huge rise of internationalized education policy in the last century (Wiseman and Baker 2005). At no time does this mean that educational policies will exactly alike for they will still be mediated by their national contexts and other nuancing factors. An influential report makes the case that there are four key institutional values at play, namely: (a) equality of opportunity and social justice; (b) development of modern individuals as a collective good; (c) dominance of academic intelligence; and (d) meritocratic achievement and education credentialism (Baker 2009). This list therefore enables us to determine in a relatively quick way how the four research studies in Part II served these core ideas/goals of schooling as shown in the last column of Table 5.1. All of them served overall purposes of promoting academic intelligence that is understandable for chapters here are devoted to developing science teaching and assessment. These are all down-to-earth, practical concerns amongst most teachers, parents, and the research community. Chapter 9, on the other hand, addressed the development of modern individuals as a collective good through its ambitious, comprehensive curriculum framework. All these chapters in this part are therefore ones that readers can read with much benefit.

References

Baker D (2009) The invisible hand of world education culture. In: Sykes G, Schneider B, Planks DN (eds) Handbook of education policy research. Routledge, New York, pp 958–968

Cochran-Smith M, Lytle SL (1999) Relationships of knowledge and practice: Teacher learning in communities. Rev Educ Res Educ 24:249–305

Darling-Hammond L, Bransford J (2005) Preparing teachers for a changing world: what teachers should learn and be able to do. Jossey-Bass, Hoboken
Deleuze G, Parnet C (2002) Dialogues II (H. Tomlinson & B. Habberjam, Trans.). Continuum
Deleuze G, Guattari F (1988) A thousand plateaus: capitalism and schizophrenia. The Athlone Press, London
Wiseman AW, Baker DP (2005) The worldwide explosion of internationalized education policy. In: Baker DP, Wiseman AW (eds) Global trends in educational policy. Elsevier, Amsterdam, pp 1–22

Chapter 6
Enhancing Science Teachers' Language Awareness with the Use of a Content-Language Integrated Framework for Developing Student Writing

Lay Hoon Seah, Jonathon Adams, and Aik-Ling Tan

6.1 Introduction

Constructing scientific texts is complex and demanding for learners as they must grapple with conceptual knowledge, epistemic practices of science, social norms related to science and the complexity of the language that is used to explain scientific phenomena (Yore 2012). It is thus not surprising that students encounter multi-faceted challenges with the use of language in science (e.g., Quílez 2019). As such, assessing student writing ought to be carried out on multiple dimensions. When teachers assess their students' science writing, science conceptual accuracy is high on teachers' agenda, given the predominant concern for conceptual development (Chua et al. 2017). However, this tendency to appraise student writing along a single dimension of conceptual accuracy can result in limited insights into students' abilities to write science and understand scientific concepts (Seah 2013).

Current frameworks available for examining student writing (e.g., Chua et al. 2017) were mainly applied by well-trained analysts for research purposes, rather than teachers for instructional purposes. This paper presents an evaluation framework for student-written scientific texts that can be used to raise science teachers' language

L. H. Seah (✉)
Social Service Research Centre, Faculty of Arts and Social Sciences, National University of Singapore, Singapore, Singapore
e-mail: lhseah@nus.edu.sg

J. Adams
Centre for Academic Persistence, Arden University, Leeds, United Kingdom

A.-L. Tan
National Institute of Education, Nanyang Technological University, Singapore, Singapore
e-mail: aikling.tan@nie.edu.sg

Y. S. Ong et al. (eds.), *A Diversity of Pathways through Science Education*,
https://doi.org/10.1007/978-981-97-2607-3_6

awareness and illustrates how the framework was trialled and adapted by science teachers as a tool to address both conceptual and language issues that students grapple with when formulating scientific texts. Specifically, this paper describes three main sections of the study: (1) the empirical derivation of the framework, (2) the use of the framework in teacher professional learning, and (3) the observed impact on the instructional practices of the teachers who utilised the framework. In focusing the research on the utility of our framework to enhance science teachers' language awareness to develop students' ability in writing scientific texts, we address the following research questions:

1. In what ways did the teachers make use of the framework in their science instruction?
2. What is the nature of the language awareness that was evident in the teachers' use of the framework in their instruction?

6.2 Literature Review

6.2.1 The Importance of the Language Lens

Language serves a constitutive role in learning science and is also a major obstacle in the process (Wellington and Osborne 2000). Not only do science learners need a minimum level of basic and academic proficiency in the language of instruction, they also need to grapple with the distinctive language features that set scientific language apart from the everyday language (Shanahan and Shanahan 2008). Learners face a variety of language-related challenges in science including technical vocabulary, everyday words with non-vernacular meanings, long noun groups, nominalisations and non-conventional ways of realizing logical reasoning (Fang 2012).

Language, as used in science, serves multiple functions, such as describing, explaining, comparing and arguing at the text level, with each genre having its own linguistic norms and requirements (Avalos et al. 2017). These texts, in turn, are bounded by unique ontological restrictions and conventions, such as 'precision', 'clarity' and 'brevity' (O'Toole 1996, p. 117). Consequently, learners face varying demands as they move between the multiple discourse communities of home and school, and when learning different school subjects. Such demands are best encapsulated by the notion of the three-language problem (i.e., home, instructional and science) proposed by Yore and Treagust (2006). Addressing the language demands of science is essential not just for English Language Learners in multilingual classrooms (the context of this study) but also for native English speakers who must grapple with the unique ways and conventions in which language is used in science. All these distinctive features of scientific language can present considerable barriers to science learning, particularly if teachers do not attend to the language needs of their students.

6.2.2 The Value of Teacher Language Awareness

Research has highlighted the important role that science teachers play in equipping students with the necessary language and literacy skills to learn and communicate scientific knowledge (Carrier 2013). However, several obstacles stand in the way for teachers, such as the lack of knowledge and skills that subject teachers need to integrate content and language learning (Cammarata and Tedick 2012) and the limited opportunities for professional learning in this area (Tedick et al. 2011). Furthermore, many content teachers believe that language and content objectives ought to be separated (e.g., Hüttner et al. 2013). Given these challenges, there is a need to raise their language awareness not only regarding the role of language in content learning (Andrews and Lin 2017) but also the language demands and the tools and skills for addressing these demands (Rutt and Mumba 2019).

Teacher Language Awareness (TLA) is a particularly useful notion for the research purpose of this study. Used mainly in second language teaching research, TLA denotes 'the knowledge that teachers have of the underlying systems of the language that enables them to teach effectively' (Thornbury 1997, p. x). It is 'essentially concerned with subject matter knowledge (specifically, knowledge of the language) and its impact upon teaching' and involves the ability to understand how the language works and to analyze it (Andrews 2007, p. 24). This study draws upon the notion of TLA by Andrews and Lin (2017), which comprises two dimensions: a declarative dimension which emphasizes the knowledge that a teacher needs and a procedural dimension which emphasizes the instructional uses of such knowledge. Within the declarative dimension, there are three components: knowledge of language (KL), knowledge about language (KAL) and knowledge about students in terms of their language knowledge and abilities (KS). The 'language' refers to relevant aspects and features of both academic language and disciplinary language that students would need to acquire concurrently to learn and communicate science (Seah and Yore 2017). This study is an attempt to demonstrate the role of the language awareness gained by the science teachers with the use of our framework.

6.2.3 Current Methods for Examining Student Writing

Most research on science writing emphasized the content or cognitive perspectives, such as the conceptual depth and accuracy of the knowledge presented (McLure et al. 2020), the intellectual quality (Chua et al. 2017), and the quality of content in arguments (Heitmann et al. 2014). Studies that attended to both the content and language aspects of student writing are comparatively less. Huerta et al. (2014) and Ruiz-Primo et al. (2004) examined students' writing in their notebooks using scoring rubrics that include criteria such as the logic of observations and predictions, inclusion of technical vocabulary, mechanics (e.g., spelling, syntax) of English language, and quality of communication. Similarly, Lee et al. (2011) examined the relations between the form (i.e., language conventions, style and organisation) and its content (i.e., scientific knowledge and understanding) in Grade three students' expository writing. The scoring rubrics used in these studies ranged from two to five levels and allowed for quantitative comparisons across students and classes. However, while the use of such rubrics provides a holistic view of the student writing, they are less useful in surfacing and pinpointing the specific language issues that students face, particularly at the word and sentence level.

At present, few evaluation methods attempt to bring together both the content and language lens in a way that could be used for professional learning of teachers to inform them of the language needs of the students. Studies such as the one by Jimenez-Silva and Gómez (2011) demonstrated that involving teachers in understanding language in science can lead to equipping teachers with the necessary skills for supporting students in science writing. This position is also adopted in the current paper, arguing that providing science teachers with knowledge of the linguistic features of science texts can lead to improvements in learner understanding through language-focused activities in the classroom.

6.2.4 SFL and Its Use in Science Education Research

Systemic functional linguistics (SFL) provides a functional-semantic view of language that allows one to link the use of linguistic components (such as noun, verb, and propositional phrases) to the meanings realised in texts and to the social and cultural contexts in which the language is used (Halliday et al. 2004). An SFL view makes explicit how meanings are realized semiotically through its linguistic components, thus allowing aspects and components of language that are distinctive and consequential in science to be uncovered (Seah 2020). Taking a functional-semantic view also means that our focus is on the extent to which the use of language fits the functions or meanings in the context of use, rather than on the accuracy of syntax based on traditional grammar rules (e.g., subject-verb agreement and tense errors).

Although SFL is useful in providing the language lens for our purpose, it has two limitations. The first limitation is the highly technical nature of SFL given its vast set of technical language for unpacking the multiple meanings embodied within individual clauses (the unit of analysis using SFL) (Bourke 2005). The entire set of technical tools can be perceived as 'very messy and complex' and 'too full and too rich for teaching and learning' (Bourke 2005, p. 93). The second limitation of SFL is that as a theoretical lens it does not provide the science disciplinary standards against which language use can be evaluated. These standards must come from the science perspective. In order for our framework to be productive (that is, in identifying the discrepancies in their writing) and practical (without overwhelming science teachers who typically do not have a language background), it needs to address these two limitations. The framework that we developed seeks to utilise a selected set of SFL analytical tools that would suffice to pinpoint the linguistic components that are problematic in student writing as well as a set of disciplinary criteria for evaluating the linguistic components against the expected way in which language is used in science.

6.3 Developing the Framework

Students' writing served as our primary source of data. The written artefacts were selected by teachers on the basis that they did not meet their curricular expectations. We analysed a total of 346 students' written responses to items from various topics collected from three classes over a period of 12 months. These items were questions or tasks that the teachers intended for either formative or summative assessment purposes. These responses were typically one- to three-sentence long answers to various kinds of questions. Examples of questions included those that requested the aim of a given experimental set-up, the relationship between variables in a table, description and/or explanation of the phenomenon that students needed to identify. Table 6.1 shows the distribution of the responses collected across the topics.

6.3.1 Context in Which the Framework Was Developed

The students were from two Grade five (age 11) and one Grade seven (age 13) classrooms in different schools. These classes, as with many classrooms in Singapore, were populated by students who came from different ethnic groups (mainly Chinese, Malay and Indian with some students from overseas) and had a range of home

Table 6.1 Range and distribution of student artefacts collected and analysed

Grade level	Topic	No. of responses
Grade 5	Changes of states and water cycle	37
	Energy	5
	Human systems	6
	Plant systems	17
	Changes of states and water cycle	94
	Energy	46
	Life cycles	22
Grade 7	Heat	68
	Physical properties of matter	12
	Separation techniques	2
	Cells	2
	Light	23
	Transport system	12

language backgrounds. While English is the language of instruction for all content subjects since Grade one, students' language experiences outside of school and at home often vary as could their language proficiency (Vaish et al. 2009). They also generally exhibited a range of science proficiency according to their teachers.

6.3.2 Analytical Approach

The development of our framework was informed by our analysis of the student artefacts. The analysis involved two phases:

1. An initial sense-making phase of analysis was conducted separately by all three researchers using our individual perspectives and expertise in science and linguistics.
2. A consolidation phase in which the multiple interpretations in the previous phase constituted the data from which we developed the framework.

6.3.2.1 The Sense-Making Phase

Student artefacts sent by the teachers were circulated among the research team for individual analysis. Our analyses were confined to student writing that were at least one clause in length. As the three of us (an SFL analyst and science education researcher (Seah), an English Language Specialist (Adams), a science teacher educator and a science education researcher (Tan)) have varied backgrounds, our analyses were thus informed by different perspectives. The team identified the errors, characterized them using a variety of perspectives, and hypothesised the possible reasons for what could have resulted in those errors. Among the perspectives that were employed in our separate analysis included:

- Checking and commenting on the student writing for conceptual accuracy (Seah and Tan),
- Examining the aims of the written activities from a language perspective to identify language issues at the word-sentence level and discourse level (Adams),
- Identifying language choices that made the student writing distinct in meaning from that expected by the teacher (Seah and Tan) and categorising them with categories adapted from SFL (Seah).

Our combined analyses afforded a more comprehensive and reliable examination of the student writing than could be expected of individual researchers working alone from one discipline. Through face-to-face and online interactions, the three researchers could also negotiate meanings and seek common agreements to understand the language challenges the students faced when writing scientific texts.

6.3.2.2 The Consolidation Phase

To make sense of our multiple analyses, we looked for patterns in our interpretations. Our interpretations were grouped into two levels, the first focusing on the meaning of each student artefact at the text (meaning of the whole response) level and the second on the word choices at the word-sentence level.

Analysis of Meaning at the Text Level. We found three categories into which we could sort the errors found in the student writing:

1. Responses with purely content errors (CE): refers to responses containing errors that could be explained by a lack of understanding of either the content or the questions that elicited the responses.
2. Responses with purely language errors (LE): refers to responses containing linguistic errors due to a lack of mastery of language use. Notwithstanding the errors, these responses suggested that the students most probably understood the question and the concepts involved.
3. Responses with content and/or language errors (CLE): refers to responses that contain a mixture of content and language errors or the errors in them could just as likely be attributed to a lack of understanding or a lack of language mastery.

Our subsequent analyses focused on students' responses belonging to categories LE and CLE. By their very nature, these responses invoked interpretations regarding the students' language choices that were of interest to us. Such interpretations are also noteworthy for their limited awareness by science teachers by virtue of their disciplinary and professional training being more focussed on content and not language.

Analysis of Word Choice at the Word-sentence Level. From our analysis of the student writing using the adapted SFL categories, we identified five of them that were useful in capturing and characterizing the language choices that the students had issues with. These categories were sufficiently distinct from each other in identifying the function of the students' word choice and comprehensive in covering all the linguistic components at the sentence level to help teachers to identify the language issues that their students encounter. Table 6.2 describes the SFL categories adapted for use in this study.

We further identified four criteria of language use that could characterize the erroneous choice of linguistic components through an iterative inductive coding process. The four criteria are 'Relevance', 'Accuracy', 'Specificity' and 'Appropriateness'. The four criteria captured the conventions and requirements of language use that were expected of students in the science lessons. These criteria could be differentiated based on the nature of the differences in meaning between the student's language choice and the range of expected responses. 'Relevance' referred to language choice that indicates either the absence of essential or the presence of unnecessary/irrelevant meaning. 'Accuracy' indicated the use of words with entirely different meanings. 'Specificity', on the other hand, denoted the use of words with a broader meaning than expected and hence lack precision of what is being referred to. Finally, 'Appropriateness' referred to the use of language with overlapping meaning that was not

Table 6.2 SFL categories

Category	Usually represented by	Function	Example
Participant	A noun or nominal group	Identifies the entity participating or involved	'the metal', 'expansion'
Process	A verb or verbal group	Identifies the action or relationship going on	'is', 'expand'
Circumstance	A prepositional phrase	Elaborates on the *Process* or one of the *Participants*	'by heat', 'on the cooler surface'
Descriptor	An adjective, adverb, comparative, or superlative	Describes the nature of either the *Process* or one of the *Participants*	'slowly', 'bigger'
Connector	A conjunction	Indicates the logical-semantic relationship between two ideas or clauses	'because', 'as'

entirely different but neither did it match what is expected. The latter usually involved the use of everyday language that had meaning that was close to the scientific perspective but yet did not share identical meaning. Table 6.3 illustrates and explains these criteria by comparing instances of the students' responses and an edited response that would be better aligned with the expected response.

Incorporating the five adapted SFL categories and the four criteria, we propose an evaluation framework that can be used to assess students' language use and characterize the erroneous language choices in them at the word-sentence level (see Fig. 6.1). Using the first letter of the four criteria, we named our framework the RASA framework.

There are three levels in this framework. Level 1 mirrors our analysis of meaning at the text level and involves selecting student writing that is suitable for subsequent analysis at the word-sentence level. This level is intended to exclude student writing in which the student writing (i) shows no flaws or either (ii) exhibits a misunderstanding of the task or (iii) exhibits a complete lack of conceptual understanding. It would not make sense to subject the last two to subsequent analysis as they are unlikely due to problems related to language use at the word-sentence level. As the preliminary selection process, Level 1 is indispensable.

Analysis at Level 2 involves identifying the specific linguistic components that deviate from the expected scientific meanings. The linguistic components would fall mainly into five main types: *Participant*, *Process*, *Descriptor*, *Circumstance*, and *Connector* (Halliday et al. 2004). Each of these types serve a distinct and integral role in construing scientific meaning (see Table 6.2). Once the linguistic component/s of the issue is identified, the analysis at Level 3 characterizes them according to one of the four criteria of relevance, accuracy, specificity and appropriateness. These criteria identify the nature of the discrepancy in meaning of the language choice at issue against alternative more suitable language choice that aligns with the scientific perspective. Levels 2 and 3 thus serve different purposes: the former identifies the

Table 6.3 The four criteria for characterizing erroneous linguistic components

Criteria for characterising linguistic component	Student's response	Edited response that is closer to the expected response	Error in student's response	Difference in meaning between the students' language choice versus expected language choice
Relevance	The temperature of water in the container increased, the time taken for water in the container to evaporate decreased	**As** the temperature of water in the container increased, the time taken for water in the container to evaporate decreased The iron ball hitting nail A have to ~~get bigger in size and~~ having a larger mass	Missing connector 'as' Redundant descriptor of the size of ball since size is irrelevant in this explanation	The student's language use either *lack* an essential linguistic component or *include* one that is redundant
Accuracy	The hot water vapour from the hot water touched the cooler surface of the ice cube and melt	The hot water vapour from the hot water touched the cooler surface of the ice cube and ~~melt~~ **condensed**	Inaccurate process	The student's language choice is *different* in scientific meaning from the expected
Specificity	Because A has more gravitational energy than B	Because **the ball in set-up A** has more gravitational energy than that in B	Non-specific noun	The student's language choice is *too broad* in meaning relative to the expected
Appropriateness	The aim is to find out if the thick celery stalk will drink more water than the thin celery stalk	The aim is to find out if the thick celery stalk will ~~drink~~ **absorb** more water than the thin celery stalk	Inappropriate process	The student's language choice *overlaps but is subtly distinct* in meaning from what is expected

grammatical function the erroneous component serves while the latter characterizes the nature of the error. If the focus is only to identify the nature of the error, Level 2 is optional. However, including this level allows one to see more patterns in terms of the language use by students.

Through the use of this multi-level framework, student writing could thus be characterized at both the meaning and word-sentence levels from an integrated content-language perspective. The limited set of metalanguage terms used within

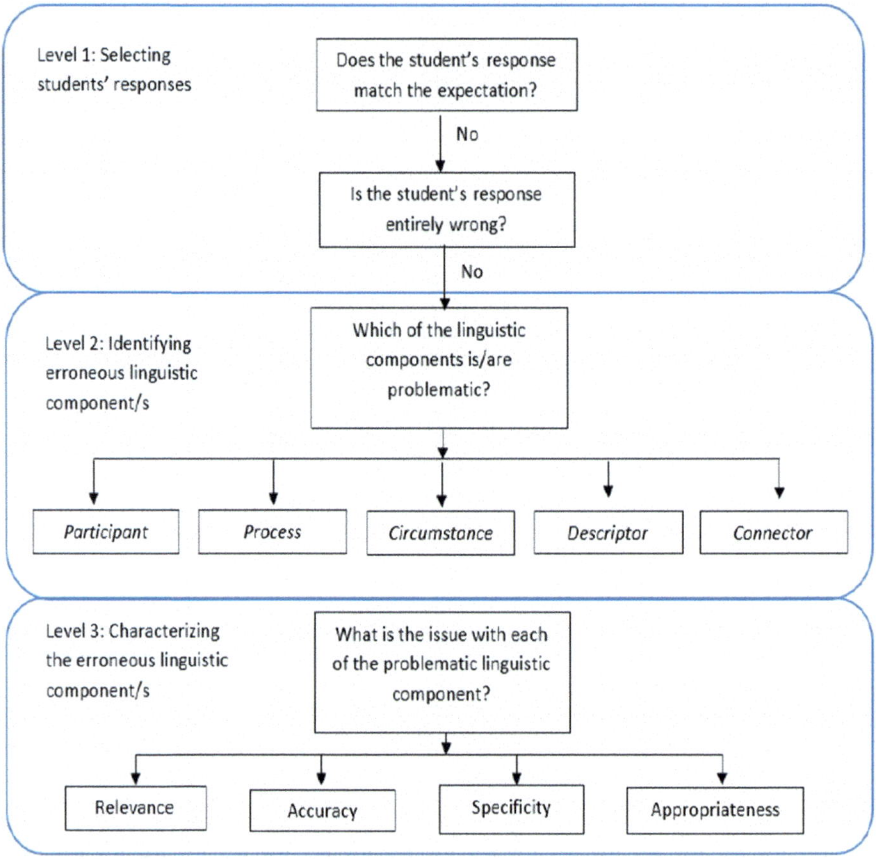

Fig. 6.1 The RASA framework

the framework also ensures that the framework is manageable and practical for use by teachers.

6.3.3 Usefulness of RASA Framework from Researchers' Experiences

We found from our analytic experiences that using the four criteria of RASA allowed us to recognize frequently occurring patterns of language use that hampered the students' capacity to represent their meanings in a clear, precise and scientifically sound way. An example was the tendency of students to use *Participants* that were not specific to the context (e.g., 'water vapour' without specifying its source). A related pattern was the omission of *Circumstances*, often resulting in incomplete responses and indicating the lack of elaboration of a certain aspect of the phenomenon being explained. Both the lack of specificity and completeness in the language choice could suggest limited conceptual understanding or unsophisticated language use that hampered the students' capacity to provide a scientifically accurate and complete response.

Such patterns of language use point to the need for more targeted pedagogical interventions that could help support students not only in better representing their meanings in scientific writing but also in enhancing their understanding of science. This is particularly important in multilingual classrooms where the primary language instruction has an emphasis on 'key words' (Tan and Saw Lan 2011). Instead of merely instructing students on particular vocabulary that is needed in scientific explanations, teachers can make use of the criteria to provide feedback on why students' choice of words is not suitable or to engage students in evaluating their peer writing. The use of the framework in Seah (2020) demonstrated the validity and reliability of the framework and its applicability in examining student writing beyond the grade levels and science topics found in the student artefacts from which it was derived.

From our previous classroom research, we observed that teachers did use some of the metalanguage (e.g., 'accurate', 'specific',) in the framework, but in ways that were ad hoc and inconsistent. This suggests that while the challenges as delineated in the framework are not unfamiliar to teachers, they need a coherent set of constructs with which they can assess students' language use more consistently. Such a consistent approach is important in enabling teachers to discern patterns of language use that are of significance in representing scientific meanings and thus the specific language needs of the students. The framework thus has value for teachers in giving feedback. First, all student written work is examined against the same criteria. Second, the feedback can be given in a systematic and organised way for the students. Third, as the focused aspects of student writing could be used systematically and as part of regular classroom practice, the teacher does not have to rewrite or restate the same points in detail repeatedly. We thus decided to trial the framework with a group of secondary science teachers to test its effectiveness in enhancing the teachers' language awareness and examine its utility (if any) in science instruction.

6.4 Applying the Framework in Teacher Professional Learning

6.4.1 Context of Professional Learning

The framework was piloted with three teachers, Baofei, Hsu-mei and Tasha (pseudonyms), teaching Grade 10 Biology classes in the same school. The school shared a similar student profile with the schools from where the student writing to develop the framework originated; the students came from multilingual backgrounds and had a range of science proficiency. These teachers had between 4 and 11 years of teaching at the time of data collection. None of these teachers had had prior in-service training or formal studies specifically around language demands in science.

The teachers participated in a researcher-practitioner collaborative inquiry (Timperley and Parr 2010) with the aim of enhancing TLA by examining their students' language challenges using student writing as the stimulus of inquiry. The inquiry process spanned four inquiry cycles. Each inquiry cycle comprised: a pre-lesson inquiry session to discuss the teacher-selected student writing and the ways to address the language issues that could be identified from the student writing; lesson observations to address these issues; and a post-lesson reflection dialogue to discuss the lesson. During each cycle, the research team supported the individual teachers to inquire into their students' challenges in writing science by identifying the gaps between student writing and the teachers' expectations based on the syllabus requirements. During the pre-inquiry session, the researchers worked with the teachers collectively. However, in the post-inquiry session, one researcher worked with each teacher to reflect on their lesson before the teachers came together to discuss with the researchers how the language challenges identified during the pre-inquiry had been addressed in the various classes, the challenges faced and the possible ways to resolve these challenges.

6.4.2 Role of Framework During Inquiry

The framework was introduced in the first inquiry session with the teachers as a tool that they could use to analyse their students' language use and to identify the specific challenges they had with particular tasks and topics. Only the specificity criterion was emphasized in this session, partly because this was the main issue with the teacher-selected student writing and it allowed us to solicit the teachers' responses to the framework. Their responses informed the ways in which the researchers made use of the framework in subsequent inquiry cycles. In the following two inquiry sessions, the student writing allowed us to illustrate more comprehensively how the four criteria could be applied to the student writing. By the third inquiry session, two of the teachers, Baofei and Hsu-mei, felt confident enough to use the framework in their classrooms. In the last inquiry session, more examples were provided to illustrate how it could be applied to analyse student writing. In addition, better differentiation between the four criteria was made and common examples of students' language use associated with each of the criteria were also provided to help the teachers better appreciate the value and utility of the framework.

The introduction of the framework provided the teacher-researcher team with the metalanguage to identify and discuss the language issues manifested in the student writing.. In the process of utilizing this framework, the teachers developed an enhanced awareness of how language worked to construct scientific meanings and how language could afford and constrain their students' construction of scientific knowledge (Seah et al. 2022). The enhanced language awareness was evident in their instructional use of the metalanguage within the framework. With a raised awareness of the challenges that their students faced, the teachers set out to address some of these challenges in their lessons, which we observed and video-recorded. For the purpose of this study, we focused only on those lessons in which the RASA framework was used in an explicit manner to address the challenges that their students were facing in the lessons.

6.5 Applying the Framework in Science Classrooms

This section sought to examine the use of the RASA framework on the teachers. This analysis was guided by the two research questions outlined in the introduction.

6.5.1 Data

The main source of data was video recordings of lessons conducted by the teachers. A total of 13 lessons of around an hour each were recorded, five from Baofei and four each from the other two teachers. The lesson videos focused mainly on the teachers and the teaching materials and captured the voices of the teachers and those of the students who contributed to the whole-class interactions. These data were supplemented by video recordings of the pre-lesson inquiry sessions and post-lesson reflection dialogues. All the video recordings were transcribed verbatim. The transcripts were analysed to identify the ways in which the teachers had made use of the RASA framework and their self-reported impact of using it (our first research question). This analysis involved analysing the transcripts for the relevant terms associated with the RASA framework (e.g., relevant, specific, appropriate) and identifying lesson episodes in which these metalanguage terms were used in a sustained and strategic manner to support students' language use. We could identify three sustained instructional activities in which the RASA framework was used. These activities represented two distinct ways in which the framework was used, in terms of their main pedagogical purpose. A third pedagogical use of the framework was derived from the post-lesson reflection dialogues in which the teachers identified ways the framework had helped them that were not apparent during the lessons. Transcripts illustrating the use and impact on the teachers are given below as evidence.

The teachers' classroom talk that invoked the use of the framework provided the data for the nature of the language awareness manifested by the teachers (our second research question). We examined the transcripts of the lessons for evidence of TLA, in particular KAL and KS, that could be inferred from the ways in which the teachers used the RASA framework. For example, the ways in which the teachers talked about the framework metalanguage were taken to indicate the teachers' KAL.

6.5.2 *Pedagogical Uses of the RASA Framework*

In this section, we present three main ways in which the teachers adopted the RASA framework as a resource to guide the analysis of student writing, and in turn to clarify the criteria for successful writing as well as provide interactive feedback in the classroom. While two of three ways were directly observed during the lessons, the third was reported by the teachers during one of the inquiry sessions.

6.5.2.1 *As a Means for Explicating Language Norms in Science*

Two of the teachers used the RASA framework in their classrooms, albeit partially, by introducing some of the metalanguage (e.g., connectors, relevance, accuracy, specificity, appropriateness) to their students as a set of tools for identifying errors in their answers. Baofei presented a list of anonymized student responses taken from their homework task in a worksheet. These responses occupied the first column of a three-column table in the worksheet. Students were required to rewrite the correct response in the second column by editing the one in the first column. The last column required students to characterize the errors using the introduced metalanguage. The teacher modelled this error correction and analysis task with several examples before getting students to do the same for the remaining anonymized student responses. He introduced the metalanguage in the process of modelling how to characterize the language issues that were present in the sample answers. Table 6.4 shows an episode in which Baofei used metalanguage terms such as 'connectors' to characterize the language issues.

In the above episode, the teacher highlighted the wrong usage of connector 'to' in the sample answer ('prevent overpopulation of lower trophic levels [idea one] to maintain biodiversity [idea two]'). The word 'to' suggested that idea one resulted in idea two when this was not the case. Instead, these two clauses should be joined by 'and' instead of 'to'. The use of metalanguage allowed the teacher to highlight the role of the connector in linking different ideas together and concurrently pointing to the correct connector that needed to be employed to indicate the accurate relationship between the two distinct ideas. In this episode, although the word 'inappropriate' was used, the emphasis was on the inaccurate use of connector and the teacher did not appear to distinguish between an inaccurate versus an inappropriate choice of words.

Table 6.4 Baofei's use of framework to highlight language issues

Speaker	Transcript
T	So what part of speech is this? It's a connecting word right? So the wrong connector is used. It's inappropriate, right? So I circle the connector. Can someone propose another connector? [Idea] One does not lead to two right? But if you put this connector in between the two, it sounds like one leads to two which is not the case. So what should be the connecting word?

Tasha did something similar as part of a post-examination review lesson. After providing a brief explanation of the meaning of the four metalanguage terms (i.e., relevance, accuracy, specificity, appropriateness), the teacher provided a catalogue of anonymised sample answers that exhibited errors within each category. Unlike Baofei who emphasized the function of the linguistic components, Tasha focused on the different ways in which the students' word choices could be enhanced by distinguishing between the different language norms in science (i.e., being 'accurate', 'appropriate' and 'specific'). The episode in Table 6.5 illustrates how she distinguished between the two norms of accuracy and appropriateness at the start of the lesson.

Tasha began by giving an example in which a student inaccurately identified the organ which produces the hormone insulin. She identified this as an 'inaccurate' word choice as the content was based on the wrong information. Her second example highlighted the inappropriate use of the words 'break down' to indicate the action of insulin on glucose. While the process does involve glucose being in a chemical reaction, the verb 'break down' did not provide the complete process in which glucose is converted into glycogen.

In both cases, the introduction of the RASA framework allowed the teachers to explicate the nature of the common language issues that students tended to display in their explanations using real-life examples. The ways they introduced the framework

Table 6.5 Tasha's distinction between language norms

Speaker	Transcript
T	Okay I give you an example of something that is not accurate that I found in the prelim paper. Some of you wrote that insulin is produced by the liver cells
S1	No by islets of langer (hans)
T	Ah, islets of langerhans are from which organ?
S	In pancreas
T	Ah in pancreas. So this is what I mean by not accurate. Now it is obviously your content. The information is inaccurate. But most of us don't make the error, most of us have this issue, where it is not appropriate, I give you an example over here. The insulin is to break down the glucose. So by the word 'break down' I understand what you are trying to say. You are trying to say the glucose is converted into something else. But the thing here is, the word break down is not appropriate. I can't say insulin breaks down the glucose, what should I say?
S2	Convert
S3	Insulin stimulates the conversion of (glucose)
T	Did you hear what she said? She said that insulin stimulates the conversion of what to what?
S4	Glucose into glycogen
T	Glucose into glycogen. So you use the same type of phrasing (if you're talking about) glucagon. Glucagon stimulates the conversion of glycogen to glucose. This is what I mean by it being not appropriate

Teacher/Peer feedback & suggest to improve on:

	Specificity	Relevance	Accuracy	Appropriateness
Pls tick (✓)				
Reason(s)				

Fig. 6.2 Table for students to conduct peer-evaluation

exemplified how the framework could be employed to raise critical awareness of how language is used in science and the conventions and requirements of its use.

6.5.2.2 As a Peer and Self-Evaluation Tool for Students

Instead of identifying sample responses for students to edit and characterize, Hsu-mei had her students exchange their written work and employ the RASA framework to peer-evaluate their partner's work after introducing and explaining the four criteria of the RASA framework. The students had to fill in a table at the bottom of their peer's explanations (see Fig. 6.2).

The table set out the criteria that the students needed to evaluate their peer's work. It also required them to explain their feedback to their peers. One observation made by the teacher was that her students seemed to make sense of and adopt the RASA framework metalanguage much more quickly than she did when first introduced to it.

6.5.2.3 As Written Feedback to Students

The evidence for the written feedback came from the self-reported description by one of the teachers. When she was asked during an interview whether there had been any changes to her teaching practices. Tasha described how her written feedback was different from the past:

> I think the feedback we give definitely is richer now ... normally we just put … the mountain symbol, it just means that you need to say more… Or we will just say that it's not specific or … "poor phrasing". When I [marked the] prelim paper...I never write a single poor phrasing... They know that it's poor but what is wrong?... So now... I can say that this is an inaccurate representation or this information is irrelevant... we are clearer in identifying their errors and we are able to point out to them.

It appears from the quote above that Tasha felt better equipped in providing more precise feedback through the use of the metalanguage terms provided in the RASA framework. Baofei, related an incident when he was helping his student one-to-one after school. He was able to deconstruct the explanation, identify the missing element in the student's explanation and advised his student accordingly. He concluded that

"So at least I'm now able to suggest an alternative method, rather than just say it's wrong, or write down what is correct".

6.5.3 Demonstration of the Teachers' Language Awareness

The multiple and varied applications of the framework were a manifestation of the procedural dimension of the teachers' language awareness. Not only did the teachers make use of the framework to analyse their students' writing, they transformed their understanding of the framework into instructional strategies that made the metalanguage terms accessible to students. The different ways the teachers made use of the framework suggested that they interpreted the purposes of the framework differently. These differences in interpretation suggest the influential role of the individual teachers' pedagogical knowledge and its integration with the declarative dimension of their language awareness (e.g., KAL and KS) in shaping the diverse application of the framework. It is also this integration that made possible the changes to their instructional practices and from which their TLA could be manifested.

However, the teachers exhibited differences in their development and manifestation of KAL. This was demonstrated in their selective use of metalanguage terms found within the RASA framework. For example, Baofei focused on Level 2 (see Fig. 6.1) by identifying the linguistic components that were problematic and, in the process, explicated the role of the various types of linguistic components in construing scientific meanings. Both Tasha and Hsu-mei focused on Level 3 of the framework by highlighting how students' words choice could be problematic. This involved elucidating the conventions and norms of science writing that could be subtle but nonetheless make a difference to scientific meanings. In making use of the metalanguage to make explicit the conceptual and language demands of science writing, the teachers thus displayed their individual KAL in terms of their understanding of the meaning and affordances of the metalanguage. Nonetheless, the nuanced differences of some of the metalanguage terms (e.g., accuracy versus appropriateness) could be more clearly expounded by them. This suggests that there is scope for developing the teachers' language awareness further so future studies are needed to understand teachers' usage of the metalanguage of the framework over time. Nonetheless, their willingness to adopt and adapt the metalanguage for their instructional purpose indicates that they saw value in using the RASA framework to evaluate their students' writing and use the student writing to discuss the language issues. The teachers' feedback on the language issues in their students' writing also demonstrated their KS, particularly the specific challenges faced by their students in their language use. This development of their KS was in turn made possible by their acquisition of KAL, which afforded the teachers a new lens to evaluate their students' writing.

6.6 Discussion

In this study, we have described how the RASA framework was developed empirically from student writing taken from their usual classroom work and assessment. The latter confers ecological validity to the framework in its use to unpack the challenges encountered by the students. We also outlined how the framework was used in our collaborative inquiry with the teachers as a lens for examining students' language use from an integrated content-language perspective. Their multiple and varied classroom applications of the framework testified to the usefulness of our framework in enhancing the teachers' language awareness.

Teachers' language awareness can be important in changing their beliefs that language and science objectives ought to be kept separate (Hüttner et al. 2013). This change in beliefs will hopefully motivate science teachers, who may be otherwise averse to the need to pay attention to language issues in their instruction to think about how they can better support students in developing their language and literacy skills in science. An understanding of the demands that language imposes on science learning is crucial for effective instructional planning and lesson enactment. With a better understanding of the language choices and patterns of use made by students, teachers can develop more targeted tasks and activities to address or discuss language choices. Teachers can also be encouraged to reflect on what factors, including their instruction, could have contributed to the challenges encountered by their students.

Different topics may impose varying demands on students' capacities to use language. Student needs also evolve over time. Given such shifting, and at times unpredictable needs of students, an openness to what is required to support students is demanded of teachers. The ability to recognize and unpack students' needs continuously is important if teachers are to constantly adjust and adapt their instruction to meet their students' needs in a timely and contingent manner (Seah and Chan 2020). The framework discussed in this study empowers teachers to continuously learn about their students and be motivated to explore alternative ways to address their students' challenges. Such a professional development approach can complement other approaches that also seek to equip science teachers with the knowledge and skills for integrated content and language instruction such as Lee and Buxton (2013). Existing approaches with their emphasis on pedagogical skills and disciplinary linguistic knowledge remain an important means to develop teachers' capacity as not all language demands of science can be inferred from students' writings alone.

Teachers' content awareness may also be enhanced when applying the framework. Any ambiguity that arises in the process of determining the issues concerning their students' language choices may encourage them to understand the scientific content and practices (e.g., the purpose of explaining and arguing) at a deeper level and go beyond the use of specialised vocabulary as a means to evaluate the correctness of students' explanations (Honig 2010). This is especially relevant for primary science teachers whose non-science-based educational backgrounds and generalist teacher training might mean that they may not always have a deep conceptual understanding

(Bowen and Bencze 2009). In this context, teachers with limited content knowledge may find it especially useful to use the framework in collaboration with their colleagues.

The way the teachers shared selected metalanguage within the RASA framework with their students suggests that they were interested not only in using the framework to understand the challenges that their students encountered in their writing but also as a tool for clarifying the criteria for writing and providing feedback to students. By getting their students to adopt the metalanguage in either self- or peer-evaluation, the teachers enhanced the possibility that the students would engage in self-regulating meta-cognitive strategies that would guide their internal dialogue when writing (Darling-Hammond et al. 2020).

Familiarising teachers with the framework for assessment and instructional purposes, however, would require professional development that shares not just how the framework is operationalised but also its affordances and limitations. In terms of affordances, the framework can develop teachers' understanding of the form-function relations of scientific language and how subtle language choices can bring about a variation in meaning. However, it is also important for teachers to recognize that the framework is not a panacea for understanding all the language challenges that students face. While the framework is useful in identifying specific issues in the students' writings at the meaning, word and sentence level, it does not tell us about students' challenges when constructing extended texts, such as the need for coherency and logical order. The framework also does not reveal the language challenges that the students faced in interpreting scientific language (e.g., when reading or listening). The framework is also not deterministic of the meaning intended by the writer. When interpreting the meaning of the students' writings, we can only base it on our literal interpretation and perhaps the teachers' best guess of what the students could have meant based on their knowledge of them through their daily interactions. To determine what students mean by their writings, it is necessary to probe them further by asking them more questions personally or getting them to use other semiotic resources like drawing (Yeo et al. 2020).

Though the study is limited in its scope with only three teachers demonstrating the usefulness of the framework, the results of this study provide support for scaling up the use of the framework in developing language awareness for in-service teachers. Future studies could investigate differences in teacher learning and the impact of adopting the framework among teachers of varied profiles as well as the trajectory of their development of language awareness over time.

6.7 Note to Our Future Colleagues

Dear colleagues,

Whether you are a human or a humanoid robot, we would like to congratulate you on the significant breakthrough that has been made with your latest invention. This invention not only saves precious time for science teachers from the tedious process of assessing their students' language use in the virtual and physical classrooms but also provides invaluable insights into the students' level of conceptual understanding and proficiency in scientific language. The artificial intelligence incorporated into the invention has skillfully mastered and integrated various frameworks (including RASA) for analysing student language. We can now better understand language use not only in student writing but also in their speech. Real-time feedback is made available to teachers on the current performance level of students as well as the possible instructional strategies and resources that can be taken to build on the former. With such a powerful assessment tool on hand, teachers would no doubt be able to leverage the illuminating qualitative feedback to design lessons to target the identified challenges that students face in their language use. Teachers' efficiency and efficacy in enhancing their students' disciplinary literacy skills and learning in a timely manner are thus enhanced.

More importantly, we hope that the invention gives fewer reasons for students to rely solely on external aids, such as artificial intelligence, in generating texts. The art of verbalizing one's own thoughts is a distinctive human trait that is worth cultivating even if there are now more tools available for text creation that depend on limited human input. No doubt learning the craft of writing can be a daunting task but the "hidden" benefits of the thinking processes (e.g., engaging in creative imagination, critical reasoning) that occur alongside the writing process are what make us distinctively human. With your new invention, students can engage in more self-directed and continuous learning to improve their oral and writing skills while keeping up with the ever-growing digital literacy knowledge and skills that are essential and ubiquitous of the society that they are in. Hopefully, with time and efforts, students learn to appreciate the freedom and power that independent thoughts and the skills to communicating them can endow on them, advantages that would only grow in importance as technology replaces ever more functions that are in the past achievable only by humans. As we continue to celebrate our successes in technological breakthroughs, let us continue to bask in the joy of writing to express our ideas, thoughts and ideals, both in science and our daily lives.

Once again, our sincere congratulations and may more breakthroughs be attained in your future research!

Sincerely,

From your happily retired colleagues

08 July 2050

References

Andrews S (2007) Teacher language awareness. Cambridge University Press, Cambridge

Andrews SJ, Lin AMY (2017) Language awareness and teacher development. In: Garrett P, Cots JM (eds) The Routledge handbook of language awareness. Routledge, London, UK, pp 57–74

Avalos MA, Secada WG, Zisselsberger MG, Gort M (2017) Hey! today i will tell you about the water cycle!": variations of language and organizational features in third-grade science explanation writing. Elementary School J 118(1):149–176. https://doi.org/10.1086/692987

Bourke JM (2005) The grammar we teach. Reflections English Lang Teach 4:85–97
Bowen GM, Bencze JL (2009) Engaging preservice secondary science teachers with inquiry activities. In: Roth W-M, Tobin K (eds) The world of science education: handbook of research in North America. Sense Publishers, Rotterdam, pp 587–609
Cammarata L, Tedick DJ (2012) Balancing content and language in instruction: the experience of immersion teachers. Mod Lang J 96(2):251–269. https://doi.org/10.1111/j.1540-4781.2012.01330.x
Carrier SJ (2013) Elementary preservice teachers' science vocabulary: knowledge and application. J Sci Teacher Educ 24(2):405–425. https://doi.org/10.1007/s10972-012-9270-7
Chua H-L, Lee S-H, Fulmer G (2017) Action research on the effect of descriptive and evaluative feedback order on student learning in a specialized mathematics and science secondary school. Asia-Pacific Sci Educ 3(4):1–22
Darling-Hammond L, Flook L, Cook-Harvey C, Barron B, Osher D (2020) Implications for educational practice of the science of learning and development. Appl Dev Sci 24(2):97–140. https://doi.org/10.1080/10888691.2018.1537791
Fang Z (2012) The challenges of reading disciplinary texts. In: Jetton TL, Shanahan C (eds) Adolescent literacy in the academic disciplines: general principles and practical strategies. Guilford Publications, pp 34–68
Halliday MAK, Matthiessen CMIM, Halliday M, Matthiessen C (2004) An introduction to functional grammar (3rd ed.). Routledge. https://doi.org/10.4324/9780203783771
Heitmann P, Hecht M, Schwanewedel J, Schipolowski S (2014) Students' argumentative writing skills in science and first-language education: commonalities and differences. Int J Sci Educ 36(18):3148–3170. https://doi.org/10.1080/09500693.2014.962644
Honig SL (2010) A framework for supporting scientific language in primary grades. Read Teach 64(1):23–32. https://doi.org/10.1598/RT.64.1.3
Huerta M, Lara-Alecio R, Tong F, Irby BJ (2014) Developing and validating a science notebook rubric for fifth-grade non-mainstream students. Int J Sci Educ 36(11):1849–1870. https://doi.org/10.1080/09500693.2013.879623
Hüttner J, Dalton-Puffer C, Smit U (2013) The power of beliefs: lay theories and their influence on the implementation of CLIL programmes. Int J Biling Educ Biling 16(3):267–284
Jimenez-Silva M, Gómez CL (2011) Developing language skills in science classrooms. Sci Act Classroom Proj Curriculum Ideas 48(1):23–28
Lee O, Buxton C (2013) Integrating science and English proficiency for English language learners. Theor Pract 52:36–42
Lee O, Penfield RD, Buxton CA (2011) Relationship between "form" and "content" in science writing among English language learners. Teach Coll Rec 113(7):1401–1434
McLure F, Won M, Treagust DF (2020) Students' understanding of the emergent processes of natural selection: the need for ontological conceptual change. Int J Sci Educ 42(9):1485–1502. https://doi.org/10.1080/09500693.2020.1767315
O'Toole M (1996) Science, schools, children and books: exploring the classroom interface between science and language. Stud Sci Educ 28:113–143
Quílez J (2019) A categorisation of the terminological sources of student difficulties when learning chemistry. Stud Sci Educ 55(2):121–167. https://doi.org/10.1080/03057267.2019.1694792
Ruiz-Primo MA, Li M, Ayala C, Shavelson RJ (2004) Evaluating students' science notebooks as an assessment tool. Int J Sci Educ 26(12):1477–1506. https://doi.org/10.1080/0950069042000177299
Rutt AA, Mumba FM (2020) Developing secondary pre-service science teachers' instructional planning abilities for language- and literacy-integrated science instruction in linguistically diverse classrooms. J Sci Teacher Educ 31(8):841–868. https://doi.org/10.1080/1046560X.2020.1760431
Seah LH (2013) Interpreting students' writings: misconception or misrepresentation? Sch Sci Rev 94(349):109–115

Seah LH (2020) What student language reveals about the demands of learning the human circulatory system. Res Sci Educ. https://doi.org/10.1007/s11165-020-09915-z
Seah LH, Chan KKH (2020) A case study of a science teacher's knowledge of students in relation to addressing the language demands of science. Int J Sci Math Educ. https://doi.org/10.1007/s10763-019-10049-6
Seah LH, Yore LD (2017) The roles of teachers' science talk in revealing language demands within diverse elementary school classrooms: a study of teaching heat and temperature in Singapore. Int J Sci Educ 39(2):135–157. https://doi.org/10.1080/09500693.2016.1270477
Seah LH, Tan AL, Adams J (2022) Developing content teachers' language awareness through practitioner-researcher inquiry into student writing. Teach Teach Educ 119:103858. https://doi.org/10.1016/j.tate.2022.103858
Shanahan T, Shanahan C (2008) Teaching disciplinary literacy to adolescents: rethinking content-area literacy. Harv Educ Rev 78(1):40–59
Tan M, Saw Lan O (2011) Teaching mathematics and science in English in Malaysian classrooms: the impact of teacher beliefs on classroom practices and student learning. J Engl Acad Purp 10(1):5–18. https://doi.org/10.1016/j.jeap.2010.11.001
Tedick DJ, Christian D, Fortune T (2011) The future of immersion education: an invitation to 'dwell in possibility.' In: Tedick DJ, Christian D, Fortune TW (eds) Immersion education: practices, policies, possibilities. Multilingual Matters, Bristol, UK, pp 1–10
Thornbury S (1997) About language. Cambridge University Press, Cambridge
Timperley H, Parr J (2010) Evidence, inquiry and standards. In: Timperley H, Parr J (eds) Weaving evidence, inquiry and standards to build better schools. NZCER Press, Wellington
Vaish V, Jamaludeen A, Roslan M (2009) The sociolinguistic survey of Singapore 2006: findings and policy implications. Research brief. National Institute of Singapore, Singapore
Wellington J, Osbourne J (2001) Language and literacy in science education. Open University Press, Buckingham and Philadelphia
Yeo J, Lim E, Tan KCD, Ong YS (2020) The efficacy of an image-to-writing approach to learning abstract scientific concepts: temperature and heat. Int J Sci Math Educ. https://doi.org/10.1007/s10763-019-10026-z
Yore LD, Treagust DF (2006) Current realities and future possibilities: language and science literacy—empowering research and informing instruction. Int J Sci Educ 28(2):291–314
Yore LD (2012) Science literacy for all: more than a slogan, logo, or rally flag! In: Tan KCD, Kim M (eds) Issues and challenges in science education research: moving forward. Springer, pp 5–23

Chapter 7
The Cognitive Demands of Secondary Science Assessment Items: Refinements to a Classification Based on Semantic Gravity and Density

Ning Charlotte Seah, Yew-Jin Lee, and Yann Shiou Ong

7.1 Introduction

All educators, not just in science education, have struggled with evaluating the cognitive demands of assessment items in valid and reliable ways that are also easy to use (Schneider 2014; Porter 2002; Penuel and Shepard 2016; Waddington et al. 2007). These endeavours have led to the widespread reliance on frameworks such as revised Bloom's Taxonomy (RBT), Webb's Depth of Knowledge, and the SOLO Taxonomy among many others (see Anderson et al. 2001). It goes without saying that there have always been disadvantages, criticisms, and trade-offs associated with any method of classification even as there are also vocal advocates to be found for each. Given the interest and practical significance of such categorisation schemes for science teaching, it is expected that research efforts here will be sustained in the near and distant future. Indeed, the stakes are high for governments and international organisations too: The highly influential OECD Programme for International Student Assessment (PISA) in 2018 administered to 15-year-olds from 79 regions and economies adapted Webb's Depth of Knowledge to determine the cognitive demands of their pool of science questions (OECD 2019).

This paper adds to the literature by refining a complimentary coding scheme derived from Legitimation Code Theory (LCT), specifically from its Semantics dimension (Maton 2014; Maton et al. 2017). While developed only within the last two decades, LCT—specifically its Semantics dimension–has recently been utilised

N. C. Seah
College of Humanities, Arts and Social Sciences, Nanyang Technological University, Nanyang, Singapore

Y.-J. Lee (✉) · Y. S. Ong
National Institute of Education, Nanyang Technological University, Nanyang, Singapore
e-mail: yewjin.lee@nie.edu.sg

Y. S. Ong et al. (eds.), *A Diversity of Pathways through Science Education*,
https://doi.org/10.1007/978-981-97-2607-3_7

to assist educators in determining the cognitive demands of science assessment items based on their levels of (i) context-dependency (from semantic gravity [SG] codes), and (ii) complexity (from semantic density [SD] codes) of ideal/expected student responses. Our proposal has expanded in a major way Rootman-le Grange and Blackie's (2018) seminal Semantics coding of undergraduate chemistry examination items by creating a more flexible classification that can also be applied across various science disciplines at secondary school levels. Indeed, we are confident that our revised coding scheme is potentially applicable for similar test formats in science at other educational levels such as in primary, middle-school, senior high, and university courses. Hence, we believe that it is a worthwhile addition to the suite of existing frameworks to gauge the cognitive or intellectual demands of assessment questions in science. In the following sections, we describe the theories and concepts behind LCT and Semantics before justifying with concrete exemplars how we reconceptualised this method with four levels of context-dependency and complexity respectively for categorising secondary science assessment items.

7.2 Theoretical Framework

7.2.1 *Legitimation Code Theory and Semantics*

Legitimation Code Theory (LCT) is an evolving theoretical framework used to analyse the process of knowledge building within social practices, that is, what kinds of talk, meanings, actions or behaviours (i.e. codes) are regarded as correct, appropriate, or legitimate for a specific social context (Maton 2014). From its roots in sociology and Systemic Functional Linguistics, LCT has now been applied beyond education to enquire how learning and expertise develops in other social fields such as law, architecture, business, and the fine arts (Martin et al. 2020). Research in LCT is moreover motivated by social justice principles due to its fundamental concern with an individual's possession as well as epistemic access to important knowledge in a community of practice (Maton et al. 2017). This access is dependent upon knowing or being able to perform different kinds of legitimation codes, that is, what is deemed to be correct, proper or acceptable within a certain social practice. At present, theory-building with LCT comprises three main dimensions—Specialization, Autonomy, and Semantics [see LCT (n.d.) for more information]—but only the Semantics dimension will be discussed here as it directly involves the classification and organisation of knowledge, and is utilised in this coding scheme for science assessment items.

Within the Semantics dimension of LCT, there are two kinds of domains: Semantic gravity (SG) refers to the degree of context-dependency of meaning in ideas/concepts/practices while semantic density (SD) refers to the degree of condensation or complexity of meanings in the latter (Maton 2014). In a nutshell, meanings that code strongly for semantic gravity would imply that they are very concrete or dependent on the immediate context for sense-making. In our case, a science question with stronger semantic gravity demands a response that is more closely tied to the context presented in the item, such as making use of information described or shown in the question stem (e.g., state the boiling point of a liquid based on a given heating curve), and thus, would not be sensible or answerable beyond this narrowly-defined context. Conversely, a question of weaker semantic gravity requires a more generalised response that would apply to or draw on ideas from many different contexts/domains or sections within the syllabus (e.g., explain the effects of heating a liquid, which can theoretically involve various concepts in chemistry or physics at different degrees of depth). Thus, semantic gravity helps analysts describe the degree of context-dependency or concreteness of meanings. On the other hand, semantic density characterises the strengths or levels of complexity of meanings. A question of stronger semantic density involves more complex meanings, which students are likely to utilise more cognitive effort in unpacking them (e.g., a question that involves multiple logical reasoning phases and/or multiple problem-solving steps to generate an acceptable response). In contrast, a question of weaker semantic density involves less complex meanings; acceptable solutions can be generated through fewer logical reasoning phases or problem-solving steps.

7.2.2 Comparisons with Other Frameworks of Cognitive Demand

Important to note here is that whether an assessment item is considered to have strong or weak SG/SD does depend on which items are being compared, the contexts of comparison (e.g., the grade level), and of course a coder's background (Blackie 2014; Maton 2014). Although SG/SD codes can vary along a continuum of strengths (they are not dichotomous codes i.e. yes/no), we adapted four levels typical of many LCT research studies such as Rootman-le Grange and Blackie (2018) to classify the degrees of context-dependency and complexity respectively of science assessment items. In this respect, research using LCT allows, in fact, requires a principled translation process to adapt abstract theories/concepts to cater for the specific purposes of the research study. By so doing, this process will facilitate iterative movements between data and theory to "simultaneously adjust the theoretical framework to the data in question and, in turn, to read the data through the theoretical lens" (van Heerden 2020, p. 3). While our coding scheme builds on past LCT-based research, we have operationalized the concepts from Rootman-le Grange and Blackie (2018) more concretely and thus refined them in a major way. At this point, we briefly justify our rationales for employing LCT and Semantics as a complimentary way to code for cognitive demands of science assessment items before going into detail about the development of our coding scheme.

Unlike the popular revised Bloom's Taxonomy (RBT), this Semantics coding scheme does not oblige a predetermined range of cognitive processes and knowledge domains involved in answering a question (Anderson et al. 2001; Krathwohl 2002; Martin et al. 2020). Both SG and SD codes can thus be "fit-for-purpose" through disciplinary and/or pedagogical norms or other criteria that are deemed relevant by the researchers. In RBT, there are suggested pairings of cognitive processes and knowledge domains (Remembering: Factual, Understanding: Conceptual, Apply: Procedural) although no such expectations are necessary when using Semantics codes, which is a distinct advantage. While RBT has potentially 24 code pairings from its twin dimensions, our coding scheme is not far behind with 16 potential pairings from two types of code relations.

Another popular assessment framework is the Structure of Observed Learning Outcomes (SOLO) Taxonomy that ascertains how well learners understand a topic (Biggs and Collis 1982). Like RBT, SOLO Taxonomy assumes a linear hierarchy of levels of understanding (i.e. cognitive complexity) from unistructural to extended abstract (prestructural understanding can be discounted as this marks an absence of knowledge). What distinguishes it from RBT is that SOLO underscores the visible, observed expression or enactment of knowledge as demonstrated by learners in tasks similar to what we are also attempting to do. There is much to commend in the SOLO Taxonomy not least because it offers both a theory of teaching and of learning as well as reportedly being able to distinguish true item complexity from difficulty (Hattie and Brown 2004). On the other hand, we argue that the 16 code combinations of context-dependency and complexity when using Semantics can possibly afford a far broader range and scope to evaluate cognitive demands of assessment items than just the four levels in SOLO. There also have been some criticism of the usage of verbs deemed helpful in guiding learning in the SOLO levels; some verbs can belong to more than one level and are thus more ambiguous than they appear (Brabrand and Dahl 2009).

A simplified though more operational version of the SOLO Taxonomy is the Depth of Knowledge (DOK) (OECD 2017) framework developed by Norman Webb (1997). In DOK, cognitive demand is determined by levels of depth that consider both the content and cognitive process to complete a task from start to finish. The four levels here include: recall and reproduction, utilizing skills and/or conceptual knowledge, strategic thinking (i.e. short-term use of higher order thinking processes), and extended thinking (i.e. extended use of higher order thinking processes) (Webb's Depth of Knowledge Guide 2009). However, it is apparent that similar issues concerning utilisation of the SOLO Taxonomy applies to DOK as well. As mentioned earlier, it is true that every existing classification method of cognitive demand has their own promotors and detractors. We are therefore not claiming that our coding scheme can supplant or is far superior to other frameworks. Instead, we are suggesting that based on ideas about context-dependency and complexity, it can offer a complimentary and relatively straightforward method to assess the cognitive demands of assessment items in science education.

7.2.3 Deriving the SG/SD Coding Scheme

Educational research studies that have used Semantics have typically reported how learning can be enhanced with instruction spanning a spectrum of context-dependency/abstraction as well as complexity in a course (e.g., Georgiou 2016; Mouton and Archer 2018; Rootman-le Grange and Blackie 2020). With respect to using LCT and Semantics in examining educational assessment and cognitive demands, research here has been limited. Georgiou, Maton, and Sharma (2014) attempted to use Semantics (only semantic gravity codes) to explore the range of context-dependency in students' responses in a university physics course that Steenkamp et al. (2019) later followed. More recently, Lee and Wan (2022) have used Semantics as a means of classifying how abstract or concrete (i.e. its cognitive demands) are science learning outcomes in the intended curriculum in schools. To assess the cognitive demands of science assessment items, this present study adapted and extended earlier work by Rootman-le Grange and Blackie (2018), which was a seminal assessment-focused study using LCT and Semantics.

In parallel with Rootman-le Grange and Blackie (2018) and Lee and Wan (2022), semantic gravity (SG) and semantic density (SD) are differentiated into four strengths/levels ranging from SG−− and SD−− being the lowest to SG++ and SD++ being the highest strengths respectively. Thus, an item coded as SG++ is more context-dependent and demands a response that is closely tied to the context presented in the item whereas an SG−− item is far less dependent on the question context. An SD++ item involves unpacking of complex meanings and demands more logical reasoning phase and/or multiple steps to generate a suitable explanation or solution as an appropriate response whereas an SD−− item does not even require relevant domain knowledge to answer it well.

Again, it is critical to note that the actual coding of items with Semantics may vary according to the context such as the grade level of the assessment items. For instance, the term "energy" may elicit more concrete or less complex acceptable responses for lower grades while graduate courses in physics may expect the same term to encompass many nuanced meanings that link multiple concepts in/across the discipline. As such, all SG and SD codes are "neither definitional nor definitive" and depend on the study context (Maton 2019, p. 3). While it may seem frustrating in that there appears to be no fixed points of reference, it may be seen as providing room to manoeuvre in the iterative process of specifying the various levels of coding for teachers.

Our proposed classification scheme targets a specific format of science assessment item common at the secondary level (from Grades 7 to 9/10) around the world, namely, those that are variously called restricted, structured, short-answer completion/supply type items. Belonging to the larger family of constructed-response test formats, we focused on items that only required brief answers (words, phrases, few sentences) that rarely exceeded four marks for each question (Frey 2018). We analysed the viability of our coding scheme based on published science questions drawn from past Singapore-Cambridge General Certificate of Education (GCE) O-level examinations across the traditional disciplines of physics, chemistry and biology. The O-level examinations are taken by 15–16 year olds at the end of Grade 10, which is typically the terminal year of secondary education in Singapore. Note that even though a question in the GCE O-level examination will have multiple sub-questions within it (e.g., Question 1a, 1b, 1c), we coded all these sub-questions as individual items. This allowed us to analyse the specific cognitive demand of each sub-question as the entire question may code for diverse and contrasting Semantics levels. Based on this approach, we therefore sampled a total of 55 items (18 physics, 23 chemistry, 14 biology) published across the years 2014–2017.

7.2.4 Deriving the Four Levels of SG Codes for Context-Dependency

We used Rootman-le Grange and Blackie's (2018) coding scheme as a basis of our work, but it is important to highlight some major differences between them. As mentioned, the earlier authors set to assess the quality of assessment items to engage meaningful learning of university chemistry whereas we wished to evaluate the cognitive demands of items across the sciences for secondary school students. They sampled 44 assessment items from one examination paper while we based our coding scheme on 55 GCE O-level items from 2013 to 2017 (4 years). Lastly, our coding scheme attempted more detailed and elaborated criteria that considered the potential awarded marks for questions coded in the different levels as well as associated keywords/command verbs that may appear in the questions for the SD coding scheme. We thus sought to be more precise in the descriptions for the expected responses detailing the various code strengths to improve decision-making. For instance, the description for SD+ in Rootman-le Grange and Blackie's (2018) coding scheme reads, "The given information needs to be manipulated—unpacked before it can be interpreted." Such a description is likely to be open to the interpretation of the coder as it is not clearly defined what the manipulation of the given information entails. In contrast, while our coding scheme for SD+ (explained later) also requires the processing or manipulation of given information, we expanded on this idea by stating that students are expected to "recall formulae/equation/knowledge and interpret/explain values/structures/information from question in multiple steps" and that "it is easy to derive the requirements of the question straightaway".

Table 7.1 below shows the four strengths of SG codes for context-dependency that we refined based on ideal/expected responses to secondary science assessment items (some of which are hypothetical), which we now explain in turn. Questions coded as SG−− are the most abstract and require responses that embody knowledge from multiple disciplines not solely limited to science in order to give a full, comprehensive answer. For example, if asked to explain the effects of heat on a metal, a complete response should include ideas about chemical and physical effects with considerations for various states of the metal that will draw on a large spectrum of concepts in the physical sciences. Next, questions coded as SG− require responses from only one major scientific discipline, but from more than one sub-section of the official syllabus document. For example, a question on comparing how heat versus electric current is transferred through a metal would require drawing on more than one sub-section of the physics syllabus because responses should include descriptions of thermal conduction utilising the particulate model of matter and the movement of charged particles under the influence of electric fields.

We acknowledge that it may be difficult to distinguish topics in the Physical Sciences as belonging strictly to chemistry or physics, but most contemporary educational systems would have written documents that state the topical boundaries within a discipline. We further recognise that different educational systems might categorise topics in very different ways, but this approach of relying on a standard/official syllabus document can offer some degree of coding consistency at least within one educational jurisdiction/district/school regardless how its science topics have been categorised. An issue might also arise when dealing with integrated science; the latter is widespread at primary and middle-school levels and consists of a few science disciplines mixed within one school subject. As such, it has sections and sub-sections that might follow arbitrary themes or grade and discipline divisions defined by respective educational institutions, which means that the SG coding for integrated science will look different from single discipline science tests. The coding scheme might also differ from country to country because of how their curriculum has been organised. Nonetheless, in all situations the SG coding will remain internally consistent for that particular context with our coding scheme.

Both SG+ and SG++ code for question responses that involve conceptual knowledge coming from a single sub-section of the syllabus. What distinguishes SG+ from SG++ questions is the extent to which contextual information in the question informs an acceptable response. Responses to SG+ questions typically require less reliance of contextual information from the question and instead, expects students to recall and apply previously learnt conceptual knowledge. For instance, the question "explain how heat is transferred through a metal rod" situates the explanation within the section of thermal physics. The context in this question ("a metal rod") provides little information (apart from the rod being made of metal) that students need to incorporate in their response. Thus, students could respond using the particulate model of matter and/or the role of delocalised electrons in the metal rod in the process of heat energy transfer. On the other hand, questions coded as SG++ strongly emphasise contextual details whereby it is mandatory for students to make use of and evaluate the information provided in the question stem. An example of such a question would be "Using the particulate model of matter, describe how does heat transfer from point A of the rod to its other end, point B." The question includes a diagram of a metal rod with a heat source at one end of the rod at point A in the same question. Consequently, students have to utilize the particulate model of matter in their answer, as requested in the question stem, and make explicit reference to points A and B of the rod in the given diagram. Thus, the expected response to the SG++ question is reliant on provided information. It should also be noted that no keywords or command verbs are highlighted for SG codes as the verbs used in

Table 7.1 SG codes for context-dependency with SG code descriptors from Rootman-le Grange and Blackie (2018) as comparison

SG code	From Rootman-le Grange and Blackie (2018)	SG code descriptors	Examples of questions
SG−−	Concepts situated in the curriculum are integrated with general everyday knowledge to create meaning that is applicable in any type of context	Scientific knowledge in the response can/must be drawn from **more than one discipline** (i.e. other disciplines of science or from other subjects)	Explain the effects of heat on a metal Explain how plants photosynthesize to create chemical energy
SG−	The question requires concepts from different sections in the curriculum to be integrated to create a unified theory that is applicable to a broader context	Scientific knowledge in the response is derived from **more than one sub-section** in the official **syllabus** document but still within one major scientific discipline	Compare how heat versus electric current is conducted through a metal rod Calculate the molecular mass of propane Explain how plants make food and how the energy generated from plants moves along a food chain
SG+	The question requires application of Chemical concept(s) from one section of the curriculum to a specific example	Scientific knowledge in the response is derived from a **specific sub-section of the official syllabus**	Explain how heat is conducted through a metal rod State one difference between alkanes and alkenes State one possible adaptation a plant may have to increase its rate of photosynthesis
SG++	The question is located in a specific section of the curriculum and only requires recall of the concepts, definitions or rules	Scientific knowledge in the response is derived from a **specific sub-section of the official syllabus** Response is highly **context dependent** i.e., requires specific information from the question stem to a significant extent	Using the particulate model of matter, describe how heat is transferred from point A to point B of the metal rod in the given diagram State the functional group of the organic compound shown in the molecular structure presented From the plant cell diagram, label which part(s) are responsible for photosynthesis and explain their functions

the question are not helpful in determining the level of context-dependency based on specificity of disciplinary knowledge drawn upon and the context-dependency of expected response. Thus, keywords are not considered for coding of context-dependency, but will be significant for the coding of complexity that we describe next.

7.2.5 Deriving the Four Levels of SD Codes for Complexity

Table 7.2 shows the descriptors for the four SD codes in our coding scheme for complexity based on ideal/expected response types by students. Science questions coded as SD−− are the least complex as they do not require any specialised domain knowledge to answer the questions. For example, a student specializing in biology or chemistry would still be able to answer a SD−− physics question adequately. If a question provides a physics formula for braking force, along with the values of the mass and acceleration of an e-scooter, a student should be able to substitute these values into the given formula in the question stem to obtain the correct braking force of the e-scooter. Hence, an average student at that grade level regardless of disciplinary background would be able to derive the answer easily given a supplied equation.

The difference in complexity between SD− and SD+/SD++ is that the latter group requires intermediate steps or more logical phases of reasoning i.e. unpacking of more condensed meanings to arrive at the final answer. However, answers for questions coded as SD− can be derived more directly since they do not require students to make any further interpretation or manipulation of the questions unlike SD+/SD++ questions. This is because there is only one scientific term/formula/structure/diagram/graph that needs to be interpreted/processed to answer the question. Another clue is that usually 1–2 marks are awarded in SD− questions and with certain command verbs such as using identify, define, or predict that are lower on the revised Bloom's Taxonomy. An example of a SD− question would be to calculate the braking force of an e-scooter given the e-scooter's mass and its acceleration. The braking force is directly derived by using Newton's Second Law of motion, which is a simple equation found in the physics syllabus.

Compared to SD+, SD++ coded questions are less straightforward and do not provide any hints on how to approach answering the question. For example, a SD+ version of the e-scooter question would require students to find the acceleration (given mass of the e-scooter) using a velocity–time graph before calculating the braking force. This question is relatively straightforward as the students can supply the answer assuming that the method has been already mastered. On the contrary, a SD++ level question would require students to first decipher the implicit demands and conceptual map of a particular question before answering it. Referring to the e-scooter example, a SD++ question could test students on the practicality of a specific speed limit on an e-scooter. The cognitive demands are now greater; students would first need to suggest a feasible mass of an e-scooter prior to calculating its acceleration using a

Table 7.2 SD codes for complexity with SD code descriptors from Rootman-le Grange and Blackie (2018) as comparison

SD code	From Rootman-le Grange and Blackie (2018)	SD code descriptors	Examples of questions
SD−−	No chemical terminology or concepts are required to answer the question	**Does not require content domain knowledge** (based on targeted "Subject Content"); students are only required to process or reason from the data/information presented Usually **1–2 mark question** Associated keywords: List, State, Plot, Recognize, Name	Using the expression: braking force = mass × deceleration, calculate the braking force acting on a 15 kg e-scooter as it decelerates at 0.9 m/s^2 Consider that the concentration of a solution is derived as the number of moles of solute divided by the total volume of solution. Calculate the concentration of aqueous sodium chloride if 0.5 mol of sodium chloride (solute) is dissolved in 100 dm^3 of water
SD−	Only one term/structure/formula is given and needs to be interpreted in order to answer the question	Using domain knowledge, only **one scientific term/formula/structure/diagram/graph** needs to be **interpreted/processed** to answer the question It is **easy to derive the requirements** of the question Usually **1–2 mark question** Associated keywords: Identify, Define, Predict	Calculate the braking force acting on a 15 kg e-scooter as it decelerates at 0.9 m/s^2 Calculate the concentration of aqueous sodium chloride if 0.5 mol of sodium chloride is dissolved in 100 dm^3 of water Write out the equation for photosynthesis in plants in symbols

(continued)

Table 7.2 (continued)

SD code	From Rootman-le Grange and Blackie (2018)	SD code descriptors	Examples of questions
SD+	The given information needs to be manipulated—unpacked before it can be interpreted	**Manipulation** of **two or more terms/structures/formulae/diagrams/graphs** based on domain knowledge required before interpretation It is **easy to derive the requirements** of the question, but multiple steps (defined as two or more phases) are required to derive the answer Involves **recalling** formula/equation/knowledge and **interpreting/explaining** values/structures/information from question in **multiple steps/**logical reasoning phases Usually **2–3 mark question** Associated keywords: Explain, Describe, Outline	Determine the braking force acting on a 15 kg e-scooter using its velocity–time graph (i.e., need to determine acceleration from graph) Calculate the concentration of aqueous sodium chloride if 50 g of sodium chloride is dissolved in 100 cm^3 of water. Express your answer in mol/dm^3 From the light intensity graph, describe how the rate of photosynthesis changes when light intensity increases?
SD++	The chemical problem must first be identified before any interpretation or manipulation can be done in order to get to a solution/answer to the question (multiple steps required)	A clear understanding of the ill-defined problem involving domain knowledge **must first be established before any interpretation** to formulate appropriate responses It is **not easy to derive the requirements** of the question, and multiple steps (defined as two or more phases) are required to derive the answer Involves **recalling** formula/equation/knowledge and **interpreting/explaining** values/structures/information from question in **multiple steps/**logical reasoning phases Usually question is **4 marks and above** Associated keywords: Calculate, Design, Analyse	Determine whether the speed limit of 25 km/h for e-scooter is scientifically reasonable Sodium chloride serves many uses in the real world, such as in nasal sprays and well drilling. Suggest why it is so versatile Explain photosynthesis and its importance in an ecosystem

suitable braking time and thus, determine its braking force. Finally, students would need to reason whether their calculations are realistic in the physical world, taking into account the safety of the rider and the surrounding road conditions. Therefore, such a question would be coded as SD++ due to its high complexity or condensation of the various scientific concepts and thinking processes involved. While SD+ and SD++ codes are close, additional clues come in the form of the number of marks that can be awarded based on the norms of the Singapore Cambridge O-level examination board as well as possible command verbs in the question stem as shown in Table 7.2.

Common to both SG and SD code descriptors, we wanted to code for and foreground features of the task in the assessment item rather than relying on inferred learner cognitive characteristics. Hence, guided by the Assessment Triangle (Pellegrino et al. 2001) and following Rootman-le Grange and Blackie (2018), our SG/SD coding avoided using the language of psychological constructs (in the Cognition vertex) that must be inferred or modelled. Instead, our codes were located at the Observation vertex to look for the performance of relevant and observable tasks/actions that learners must do to answer questions acceptably. While there is still an interaction between the learner and tasks, focusing on the nature of tasks (i.e. what must the learner be able to do) at the Observation vertex was felt to add greater objectivity and reliability in our coding scheme.

7.2.6 Illustrative Applications of the Refined SG and SD Coding Scheme

In this section, we provide worked examples of our coding scheme adapted from Rootman-le Grange and Blackie (2018) and articulate our decisions and rationales in classification. Three typical questions drawn from the major science disciplines for secondary students will be presented here. Due to copyright restrictions, these examples closely resemble official GCE O-level examination items from Singapore, but are not themselves the official questions.

a. **Physics**: An iron ring hangs from a string. A bar magnet is positioned close to the ring. The iron ring is then attracted to the bar magnet. Explain why.

The domain knowledge of this question is taken from the sub-section of "Electricity and Magnetism" in the official physics syllabus document from Singapore. In terms of context-dependency, we have coded this question as SG++ indicating that it is concrete and very context-dependent. A good response needs a student to pay close attention to the given information above the diagram as well as within the diagram itself. When answering, a student needs to notice the material of the ring as well as which pole of the magnet is facing the ring. As the question includes Fig. 7.1, whereby the bar magnet is specifically labelled with north–south (N–S) polarity, the student is expected to deduce and mention the induced poles in the ring (i.e. induced S-pole on side closer to bar magnet) based on the polarity of the side of the bar magnet closer to the ring (i.e. N-pole). Thus, a response that only mentions "ring is attracted to the bar magnet as it is made of a magnetic material" is inadequate. Hence, this question is coded as SG++ as the required response is highly dependent on the context provided in the question i.e. the ring's material and the polarity of the magnet is crucial to the formulation of a complete and correct response. For complexity, we have categorised the question as SD+ as students have to "explain" the phenomenon of magnetic attraction (c.f. associated keywords for SD+, Table 7.2) in this 2-mark question. The answer involves more than one logical reasoning phase. First, students need to state that the ring is made of a magnetic material. Second, students need to state that the side of the ring closer to the north pole of the magnet is induced as a south pole and it is attracted to the north pole of the magnet given the property that opposite magnetic poles attract.

b. **Chemistry**: The diagram below depicts an electrolysis set-up. Dilute aqueous sodium chloride forms hydrogen and oxygen through electrolysis.

This question tests content from the section of "Chemistry of Reactions" in the official chemistry syllabus document in Singapore. Prior to this sub-question, students were required to state the ionic equations for reactions occurring at the anode and cathode, essentially demonstrating the electrolysis of water. With respect to context-dependency, this sub-question is coded as SG++ as students may need to refer to the ionic equations that they have written in the previous part to show the theoretical volume ratio of hydrogen and oxygen produced in the set-up. As students are required to refer to a previous part of the question, context is important in the formulation of their answers. In terms of complexity, we have coded it as SD+ as multiple steps, including making explicit reference to their answers in the previous question part are required to answer this 2-mark question. Through correctly interpreting the ionic equations for the reactions in the anode and cathode, students can then deduce the theoretical volume ratio of hydrogen to oxygen as 2:1 (Figs. 7.2 and 7.3).

c. **Biology**: The following diagram shows a germinated pollen grain.

Fig. 7.1 Example of a physics question involving bar magnet and iron ring

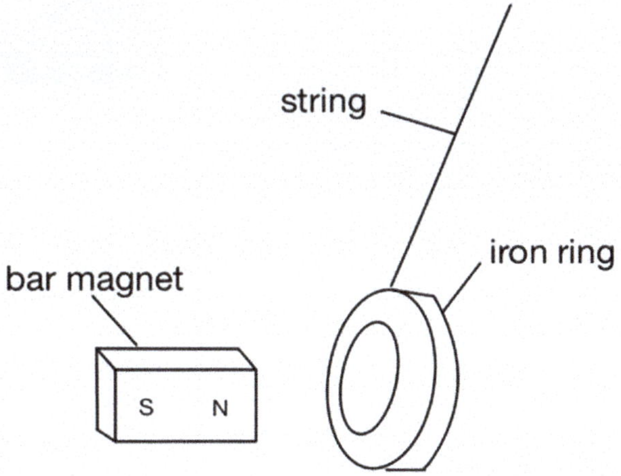

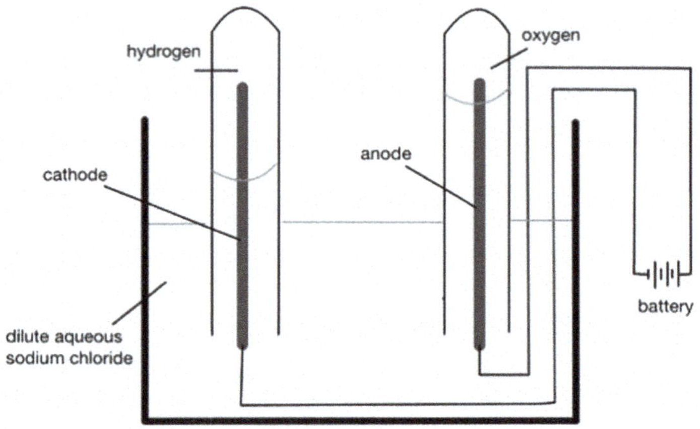

The two gases are collected and their volumes are then measured. Theoretically, the ratio of hydrogen to oxygen should be 2:1. Oxygen has a higher solubility in water than hydrogen. This causes a change in volume of the respective gas collected. Why is the theoretical ratio of hydrogen to oxygen 2:1?

Fig. 7.2 Example of a chemistry question on electrolysis of water

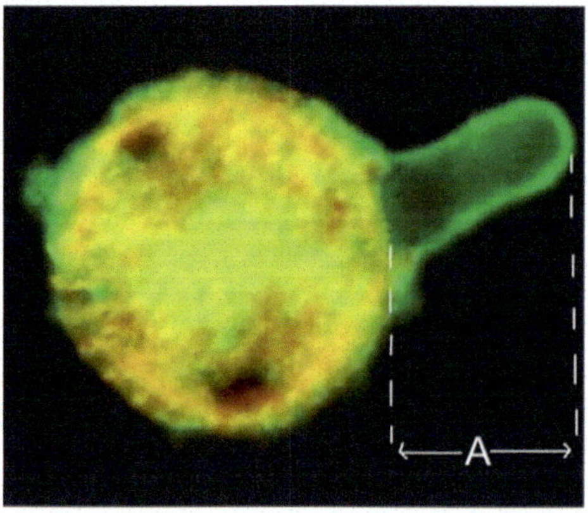

Fig. 7.3 Example of biology question on a pollen grain. *Note* From '*Germination pollen grain*' [Photograph], by AJC1 (2013), Flickr (https://www.flickr.com/photos/47353092@N00/10170683834). CC BY-SA 2.0

ci Define sexual reproduction

This sub-question references the biology syllabus section on "Maintenance and Regulation of Life Processes". For SG, this sub-question was coded as SG+ given that it is less context-dependent and only requires students to recall the definition of sexual reproduction without the need to make reference to any other information in the question. In terms of SD, we have coded the sub-question as SD+; while it is a simple 1-mark definition question (which would be SD− based on associated keyword alone), it requires several key phrases and concepts, warranting multiple logical reasoning phases.

cii. Name the part labelled A.

This sub-question was coded as SG++ because of its high context-dependency; only through examining the given diagram would students be able to identify the unknown part labelled A. It was categorised as SD– since students are simply required to identify the part labelled A without any manipulation of the information presented in this 1-mark question.

ciii. Describe **and** explain how the tip of part A reaches the female nucleus in a flower.

Similarly, we have coded the question as SG++ since students are still required to make reference to the part labelled A. It was coded as SD+ as there are multiple steps to the answer and students need to provide description along with their explanation to this 2-mark question. Both command verbs: "describe" and "explain" are associated with SD+. The requirements of the question are easily derived, however, as students have already identified the unknown part labelled A in the previous sub-question.

7.3 Empirical Analysis with the Coding Scheme

7.3.1 The Semantic Plane of SG and SD

Both SG and SD codes can be presented orthogonally on a Cartesian plane where they move independently along a continuum on a semantic plane. We can divide this plane into four quadrants to display four combinations of SG and SD code modalities as shown above in Fig. 7.4. This enables us to better visualise the meanings of coded items in terms of their context-dependency and complexity in a graphical form. Based on Fig. 7.4, Quadrant 1 depicts questions that are less context-dependent as well as more complex in nature. Quadrant 2 represents items that are also low in context-dependency (i.e., more abstract), but low in complexity. Question responses depicted in Quadrant 3 expect students' responses to be highly context-dependent (i.e., more concrete), yet not requiring any specific domain knowledge (i.e., low complexity). Finally, Quadrant 4 codes for responses that are the most context-dependent, and requires multiple logical thinking phases or problem-solving steps (i.e. high complexity) to derive appropriate responses.

Putting our coding scheme to the test, Fig. 7.5 depicts the distribution of SG/SD codes from our sample of 55 questions. Our data was therefore restricted to only Quadrants 3 and 4 with SG+ (14 items or 26%) and SG++ (39 items or 74%) codes for context-dependency, and SD– (27 items or 51%) and SD+ (26 items or 49%) codes for complexity. The majority of the questions ($n = 21$) were coded as SG++/SD+ followed closely by SG++/SD– codes ($n = 18$) while only five questions were coded as SG+/SD+. Being short-answer structured or supply type question items requiring specific answers to a question (usually based on a given stimulus), it stands to reason that this overall semantic profile was observed. These types of questions (mainly

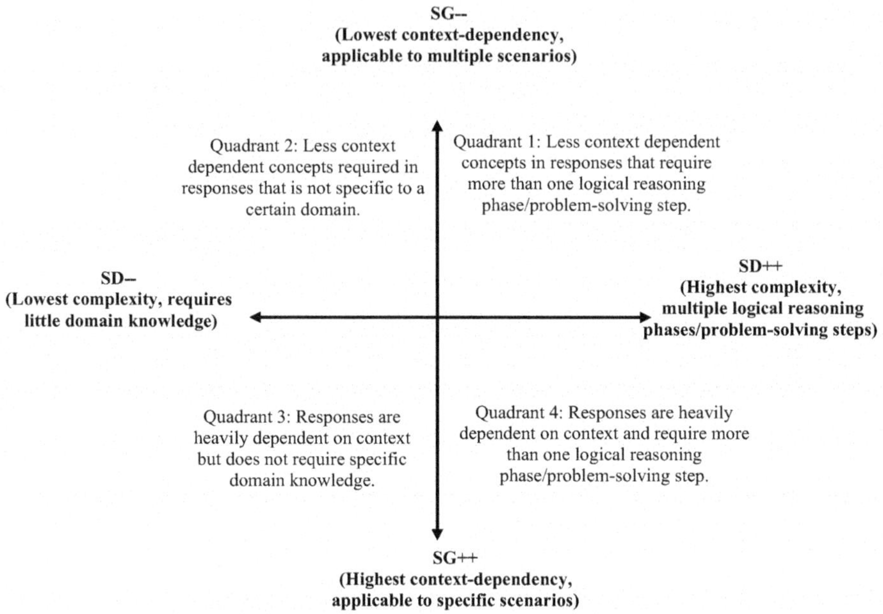

Fig. 7.4 The Semantic plane showing the four combinations of SG and SD codes in each quadrant

SG++ codes) therefore required a high degree of contextual information based on data in the question stem, which meant that students had to pay careful attention to the given "stem of information" (MOE, p. 5). Also, responses were drawn from just one specific sub-section of the syllabus with respect to these SG codes, that is, expected responses were narrowly focused in its conceptual spread. In terms of SD codes, it was also tightly clustered into two quadrants albeit achieving more balance in frequency between SD− and SD+ codes. In terms of complexity, it was therefore relatively easy to derive the requirements of the question through simple interpretation or a few steps of logical thinking/problem-solving that corresponded to a range of one to three marks for each question.

7.3.2 *Inter-Rater Reliability with Weighted Kappa*

To check the reliability of our SG/SD coding scheme, we utilized weighted kappa as a measure for inter-rater agreement for the 55 items coded. Kappa measures the proportion of agreement across multiple coders corrected for chance (Fleiss and Cohen 1973). The kappa scale ranges from −1 to +1 where a negative value implies a "poorer than chance agreement", 0 implies an agreement that is "exactly chance" and a positive value implies a "better than chance agreement" (Fleiss and Cohen 1973, p. 613). Moreover, there is an implicit assumption that all disagreements are equally

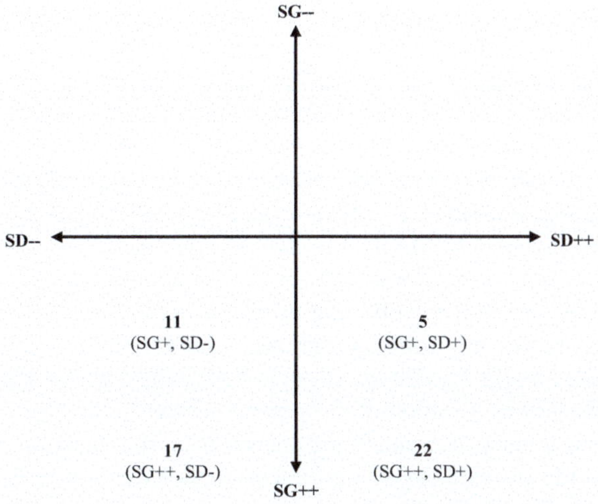

Fig. 7.5 Frequency distribution of coded GCE O-level items (n = 55) from Singapore in the three sciences in the semantic plane

significant and crucial. Since the four levels in our SG/SD coding schemes form an ordinal rating scale, we used weighted kappa so disagreements can be weighted by their magnitudes. In other words, further disagreements (e.g. question parts where one rater coded as SG++; the other as SG−−) were weighted less or indicates lesser agreement than near disagreements (e.g. question parts where one rater coded as SG+; the other as SG−). In this study, for SG codes the weighted kappa (linear weight) was 0.587 ($p < 0.001$) while for SD codes it was 0.608 ($p < 0.001$) among two coders (first and third authors). Hence, the weighted kappa values indicate moderate agreement for the SG codes and substantial agreement for SD codes (Fleiss and Cohen 1973). Our reported weighted kappa values reflect the reliability extent that can be reasonably achieved with approximately 20 h of training time.

7.4 Conclusion and Discussion

This study was primarily motivated by the longstanding search among educators for better tools to determine the cognitive demands of assessment items. We have therefore improved a complimentary coding scheme in Tables 7.1 and 7.2 from the Semantics dimension of LCT to assess science assessment items at the secondary school level. By adapting and expanding earlier work by Rootman-le Grange and Blackie (2018), we refined in a major way four levels of context-dependency and complexity based on semantic gravity and density respectively that yielded a total of 16 possible code combinations of cognitive demand based on Semantics. We are encouraged that the reliability of the coding scheme from its weighted kappa values showed moderate to substantial agreement based on our test sample of 55 test items from Singapore. The empirical analysis of these sample items in Fig. 7.5 showed that they were restricted to Quadrants 3 and 4, with the majority of items located at SG++ (highly content-dependent) while they seemed equally distributed between SD− and SD+ codes (medium complexity). As we have explained, this semantic profile seems reasonable given the nature of supply type item formats that required short answers by students.

Looking at the overall changes to the coding scheme first devised by Rootman-le Grange and Blackie (2018) in Tables 7.1 and 7.2, the differences are more pronounced with respect to SD than to the SG code descriptors. Our SD codes are more elaborate and show the various criteria in each level based the number of hypothetical steps in order to solve the problem, deployment (or not) of scientific knowledge, the use of keywords (which relies on Bloom's cognitive processes), and the allocated marks for that question. For SG codes, we paid attention now to the number of disciplines, location of the student answer in the (sub-) sections of the syllabus, and how context-dependent the answer is on the question. We believe that these criteria would provide classroom practitioners sufficient practical guidance to classify the cognitive demands of their assessment items according to the two dimensions of context-dependency and complexity in Semantics.

Our coding scheme has some potential drawbacks, which revolve around the uneven spread of codes displayed from our sample. As seen in Fig. 7.5, SG−− and SG- codes were absent in our trial coding, likewise for SD−− and SD++ codes. Nonetheless, because structured questions have very specific learning outcomes and are often based on a given stimulus, this might account for the restricted semantic profile that was observed here. Indeed, it is unlikely that these short answer supply questions would require answers that involve concepts distributed across multiple sections of the syllabus or even have multi-disciplinary answers at the O-level although such events are catered for by our coding scheme. It was also believed to be a chance phenomenon that SD−− codes were not observed during our trial coding for structured questions; these item formats were present in other O-level science questions outside of our study sample. While our trials only contained GCE O-level questions, we believe that this coding scheme is potentially applicable across other education levels as well as being relatively straightforward to implement given there are only two distinct dimensions (SG/SD) with four levels/strengths. We thus invite researchers interested in assessment to utilise and evaluate our coding scheme so as to further improve upon it. Other researchers interested in establishing relationships between item features and item difficulty could also extend the presented work in this direction.

7.5 Notes to Our Future Colleagues

The urge to categorize or rank what we humans deem as important, valuable, and meaningful has always been part of our social existence. Since time immemorial, questions have asked about who is the most beautiful, bravest, cunning, evil as well as practical questions of life regarding what is the strongest, the tastiest, the most durable and so on. Similar aspirations to classify and measure also exist within education domains with the desired attributes that we most typically value located in the cognitive domain. Certainly, we don't foresee research in this area to cease to you working in 2050! The reasons we think are simple: The work of education, for the most part, has been all about deliberating over issues of what is ability/competency and who has it. These twin, inseparable aspects of schooling have not diminished since the invention of schools writ large across world cultures. Indeed, our book chapter is no exception as we explore a different way of classifying intellectual challenges of examination test items by which one can then make valid inferences concerning "ability" in science education. We therefore feel that it is very unlikely the situation 30 years down will find anything drastic to displace this search for better definitions and measurements of competency/ability among learners.

Our message to our future colleagues is not, however, an unconditional surrender to doom because it is possible to get good education in an age of measurement (Beista 2011). We are optimistic that there will be a recognition for more 'authentic' or realistic measures of competency and knowing arising from schooling. This we hope can be coupled with greater acceptance that skill, knowledge or expertise are multi-faceted in nature and that schooling has been rewarding a narrow spectrum of human ability for too long. In the next decades, we are hopeful to dream of a broader range of desired outcomes from education such as empathy, care etc. that are valued, if not assessed in meaningful ways. In other words, there will be a wider appreciation of what constitutes intelligence and human thriving, a wider set of important skills and knowledge of educational purposes.

Moreover, we anticipate that the nature of assessment itself will be forced to undergo dramatic changes if what we read in the newspaper today (Chia 2023) is true concerning the rise of intelligent chatbots. The latter, it is said, can instantaneously compose texts and solve equations in human-like fashion thereby threatening the benefits that arise from "independent" work by students. Commentators, however, speculate that these AI-algorithms will exist alongside the process of learning (and testing), and teachers will instead be setting more assignments that test human creativity and application of knowledge in novel ways. These are the distinguishing features that separate machines from humans while others believe these AI applications are over-hyped. It now leaves you, dear colleagues, to decide if these brief thoughts concerning the end purposes of education, human learning, and assessment have been prescient or badly off-tangent, and write a book chapter about it!

Ethics Approval No ethics approval was sought as no human subjects were involved in the study.

References

AJC1 (2013) Germination pollen grain [Photograph]. Flickr. https://www.flickr.com/photos/47353092@N00/10170683834

Anderson LW, Krathwohl DR, Airasian PW, Cruikshank KA, Mayer RE et al (eds) (2001) A taxonomy for learning, teaching and assessing. A revision of Bloom's taxonomy of educational objectives. Addison-Wesley Longman, White Plains, NY

Beista G (2011) Good education in an age of measurement: ethics, politics, democracy. Routledge, New York

Biggs JB, Collis KF (1982) Evaluating the quality of learning. Academic Press, New York

Blackie ALM (2014) Creating semantic waves: using legitimation code theory as a tool to aid the teaching of chemistry. Chem Educ Res Pract 15(4):462–469. https://doi.org/10.1039/C4RP00147H

Brabrand C, Dahl B (2009) Using the SOLO taxonomy to analyze competence progression of university science curricula. Higher Educ 58:531–549

Chia O (2023) Teachers v ChatGPT: schools face new challenge in fight against plagiarism. The Straits Times. https://www.straitstimes.com/tech/teachers-v-chatgpt-schools-face-new-challenge-in-fight-against-plagiarism

Fleiss JL, Cohen J (1973) The equivalence of weighted Kappa and the intraclass correlation coefficient as measures of reliability. Educ Psychol Measur 33(3):613–619

Frey B (2018) Supply items. https://methods.sagepub.com/reference/the-sage-encyclopedia-of-educational-research-measurement-and-evaluation/i20333.xml

Georgiou H (2016) Putting physics knowledge in the hot seat: the semantics of student understandings of thermodynamics. In: Maton K, Hood S, Shay S (eds) Knowledge-building: educational studies in legitimation code theory. Routledge, Abingdon, pp 176–192

Georgiou H, Maton K, Sharma M (2014) Recovering knowledge for science education research: exploring the "Icarus Effect" in student work. Can J Sci Math Technol Educ 14(3):252–268

Hattie JAC, Brown GTL (2004) Cognitive processes in asTTle: the SOLO taxonomy. asTTle Technical Report #43. University of Auckland/Ministry of Education

Krathwohl DR (2002) A revision of bloom's taxonomy: an overview. Theory Pract 41(4):212–218. https://doi.org/10.1207/s15430421tip4104_2

LCT [Legitimation Code Theory] (n.d.) Theory. https://legitimationcodetheory.com/theory/

Lee Y-J, Wan D (2022) How complex or abstract are science learning outcomes? A novel coding scheme based on semantic density and gravity. Res Sci Educ 52(2):493–509. https://doi.org/10.1007/s11165-020-09955-5

Martin JR, Maton K, Doran YJ (2020) Accessing academic discourse: systemic functional linguistics and legitimation code theory. Routledge, Abingdon

Maton K (2014) Knowledge and knowers: towards a realist sociology of education. Routledge, London

Maton K, Hood S, Shay S (2017) Knowledge-building: educational studies in legitimation code theory. Routledge, London

Maton K (2019) Semantics from legitimation code theory: how context dependence and complexity shape academic discourse. https://legitimationcodetheory.com/wordpress/wp-content/uploads/2018/07/2019Maton-Semantics-intro.pdf

Ministry of Education [MOE] (2019) Physics. Singapore-Cambridge General Certificate of Education Ordinary Level (2021). Syllabus 6091. Singapore Examinations and Assessment Board

Mouton M, Archer E (2018) Legitimation code theory to facilitate transition from high school to first-year biology. J Biol Educ 53:2–20. https://doi.org/10.1080/00219266.2017.1420681

OECD (2019) PISA 2018 science framework. https://www.oecd-ilibrary.org/docserver/f30da688-en.pdf?expires=1625010963&id=id&accname=guest&checksum=994E1F816E11C9A91C3421AC0C48FD37

Organisation for Economic Co-operation and Development [OECD] (2017) PISA 2015 assessment and analytical framework: science, reading, mathematics, financial literacy and collaborative problem solving, revised edition, PISA, OECD Publishing, Paris. https://doi.org/10.1787/9789264281820-en

Pellegrino JW, Chudowsky N, Glaser R (eds) (2001) Knowing what students know: the science and design of educational assessment. National Academies Press, New York

Penuel WR, Shepard LA (2016) Assessment and teaching. In: Gitomer DH, Bell CA (eds) Handbook of research on teaching. American Educational Research Association, New York, pp 787–850

Porter AC (2002) Measuring the content of instruction: uses in research and practice. Educ Res 31:3–14

Rootman-le Grange I, Blackie MA (2018) Assessing assessment: in pursuit of meaningful learning. Chem Educ Res Pract 19:484–490. https://doi.org/10.1039/c7rp00191f

Rootman-le Grange I, Blackie MA (2020) Misalignments in assessments: using semantics to reveal weaknesses. In: Winberg C, McKenna S, Wilmot K (eds) Building knowledge in higher education: enhancing teaching and learning with legitimation code theory. Routledge, New York, pp 72–85

Schneider J (2014) From the ivory tower to the schoolhouse: how scholarship becomes common knowledge in education. Harvard Education Press, Cambridge

Steenkamp CM, Rootman-le Grange I, Müller-Nedebock KK (2019) Analysing assessments in introductory physics using semantic gravity: refocusing on core concepts and context-dependence. Teach Higher Educ 12:1–16. https://doi.org/10.1080/13562517.2019.1692335

van Heerden M (2020) (How) do written comments feed-forward? A translation device for developing tutors' feedback-giving literacy. Innov Educ Teach Int 58:555–564. https://doi.org/10.1080/14703297.2020.1788411

Waddington D, Nentwig P, Schanze S (eds) (2007) Making it comparable: standards in science education. Waxmann

Webb NL (1997) Criteria for alignment of expectations and assessments in mathematics and science education. Council of Chief State School Officers and National Institute for Science Education Research Monograph, National Institution for Science Education

Webb's Depth of Knowledge guide (2009) Career and technical education definitions. https://www.aps.edu/sapr/documents/resources/Webbs_DOK_Guide.pdf

Chapter 8
Formative Assessment in Primary Science Classrooms in Singapore: Teachers' Pedagogical Reasoning and Practices

Fatema Anis Hussain and Dennis Kwek

8.1 Introduction

Over the past few decades, the role of the learner in the learning process has received considerable attention with a greater focus on assessment practices, which inform students' ongoing learning in the classroom. Research in the 1990's assigned the learner the key role in the assessment process based on trends in the fields of socio-cultural learning, metacognitive and self-regulation theory (Black and Wiliam 1998; Sadler 1998; Shepard 1995; Torrance and Pryor 1998). Fundamentally, the learner needs to be able to assess and improve on the work produced through self-monitoring and self-regulation (Butler and Winne 1995). Rather than assessing the product of the students' learning, 'Assessment *for* Learning' (AfL) or formative assessment is "specifically intended to provide feedback on performance to improve and accelerate learning" (Sadler 1998, p. 77).

Formative assessment is understood in different ways, but the plural interpretations converge on the idea of teacher or student feedback, which informs ongoing student learning as well as teacher instruction. For Wiliam (2006), the most important reason to use formative assessment is that it allows for feedback to be given during instruction. Feedback is key to formative assessment, which if used appropriately can scaffold student learning (Hattie and Timperley 2007; Torrance and Pryor 1998). Formative assessment essentially comprises activities undertaken by both teachers

F. A. Hussain (✉) · D. Kwek
Centre for Research in Pedagogy and Practice, National Institute of Education, Nanyang Technological University, Nanyang, Singapore
e-mail: fatema.hussain@nie.edu.sg

D. Kwek
e-mail: dennis.kwek@nie.edu.sg

Y. S. Ong et al. (eds.), *A Diversity of Pathways through Science Education*,
https://doi.org/10.1007/978-981-97-2607-3_8

and students in assessing themselves, yielding information to be used as feedback for revision of teaching and learning (Black and Wiliam 1998).

A range of school curricular disciplines focus on inquiry, which denotes learner-centredness, knowledge construction and problem-solving (Khalaf 2018). Inquiry is central to science learning and in 2008, was identified as the guiding pedagogical framework for science education change (Kim et al. 2013). Beyond what they know, students in the inquiry classroom need to know how they know what they know, and why they know what they know (Duschl 2003). Further, students' understanding, and evidence of their understanding need to be anchored in disciplinary ideas and reasoning, as Coffey et al. (2011) observe. They argue that authentic and effective formative assessment is specific to the discipline and requires the teacher to attend to the knowledge and nature of the discipline. Inquiry-based learning in the science classroom entails the active role of the learner in constructing scientific knowledge through an understanding of discipline-based skills (Kidman and Casinader 2017).

To facilitate inquiry-based learning, formative assessment practices need to be woven into teachers' instructional repertoires (Abell and Volkmann 2006). Eliciting evidence of student learning, ascertaining where students are in their learning trajectory, and adapting and modifying instruction based on students' learning needs (Black and Wiliam 1998) involve the teachers' pedagogical reasoning or decision-making, which relates to their pedagogical content knowledge or PCK (Shulman 1987). The enactment of formative assessment practices, as Tan and Tan (2014) observe, is a skilled and complex task, and largely depends on the teachers' content knowledge and PCK as well as their understanding of the context and learners, and their belief systems. Examining science teachers' practice through the PCK lens is important (Loughran et al. 2012), and this article illustrates primary science teachers' pedagogical reasoning in relation to their formative assessment practices in Singapore classrooms. Using the lens in Hussain's (2022) study, the focus is on elucidating teachers' pedagogical reasoning both as a way of thinking and as enacted in the classroom. The article aims to unpack how teachers "navigate the messy terrain of formative assessment" (Tan and Tan 2014, p. 69) by tapping into their pre-lesson and in class teaching decisions, and how they manifest into practices, which formatively assess student learning. Accordingly, the research questions are:

1. What is the nature of formative assessment practices in primary science classrooms in Singapore?
2. What is the nature of primary science teachers' pedagogical reasoning in relation to formative assessment of student learning?

In what follows, the article initially provides a brief overview of the Singapore context. Theoretical underpinnings of formative assessment and teachers' pedagogical reasoning are highlighted followed by methodological details of the study. Subsequently, the findings are presented, which elucidate the nature of formative assessment practices and teachers' decision-making focusing on pedagogical aspects such as student learning activities, classroom talk, feedback, and visible learning (Hattie 2009). The findings are then discussed followed by the key implications.

8.2 The Singapore Context

Broadly speaking, high-stakes examinations have been one of the defining pillars of the Singapore education system since its post-independent days in the 1960s (Tan et al. 2008). Based on a large-scale study, Hogan et al. (2013) observe the "tight coupling or alignment of national high stakes assessment system and classroom instruction" (p. 60). Moreover, the routine, instructional regime stresses teacher talk and curriculum coverage to ensure examination success (Mourshed et al. 2010). Parents and students often regard classroom feedback as a tool to enhance academic success, which is seen as instrumental for social mobility (Tan and Wong 2018). Over the years, local students have performed consistently well in international benchmarking studies such as the Trends in International Mathematics and Science Study (TIMSS) and the Programme for International Student Assessment (PISA). Tan et al. (2017) observe that Singapore's highly structured, teacher-centric and content-focused education system has often been criticised for producing exam-savvy students who underperform in critical thinking or creative tasks.

To cater to the demands of a knowledge-based economy and globalisation, the Ministry of Education (MOE) has implemented wide-ranging reforms to reduce the over-emphasis on academic achievement. *Thinking Schools, Learning Nation* (TSLN) announced in 1997 describes a nation of thinking and committed citizens capable of meeting future challenges, and an education system geared to twenty-first century needs. In 2004, MOE launched *Teach Less, Learn More* (TLLM), which de-emphasises rote learning and a 'one size fits all' type of instruction, and posits experiential discovery and differentiated teaching. Since 2009, MOE has progressively implemented the recommendations of the Primary Education Review and Implementation (PERI) Committee, which encompasses a stronger focus on AfL and holistic assessment. In 2015, MOE signalled major policy shifts from teacher-centric, exam-oriented pedagogy to self-directed student learning. The new Primary School Leaving Examination (PSLE) scoring system (with effect from 2021) indicates a student's absolute performance in a grade band instead of documenting the student's performance relative to his/her peers in the cohort (Seah 2016).

Other than blunting the impact of summative assessment, MOE has continually reviewed its policy and curriculum to prioritise formative assessment and students' self-directed learning. The spirit of inquiry is central to the science curriculum in Singapore, as evident in the focus on nurturing the student as an inquirer through knowledge, skills and dispositions (MOE 2014). The syllabus posits formative assessment via feedback, which allows students to monitor their learning progress, and enables teachers to ascertain students' achievement of learning goals and the effectiveness of their instructional repertoires. AfL practices potentially facilitate inquiry-based learning thus, the focus now is on key theoretical understandings undergirding formative assessment.

8.3 Theoretical Framework

This section focuses on salient aspects of the article: formative assessment and teacher pedagogical reasoning. Feedback is a key facet of classroom assessment because it permits monitoring of students' progress toward learning goals and also informs teaching decisions (Stiggins and DuFour 2009). Teachers' pedagogical reasoning comes into play in generating and sustaining the opportunities to formatively assess student learning via the provision and uptake of feedback, and use of visible learning practices. Broadly, teachers' decision-making shapes students' engagement with tasks and classroom talk, which inform students' learning and teachers' instructional repertoires.

8.3.1 Feedback and Visible Learning

The role of feedback as a formative assessment tool hinges on the visibility of teaching and learning in the classroom, which involves making the process of teaching and learning transparent or visible to both teachers and students (Hattie 2009). Communicating goals and standards is a key element of visible learning. To provide effective feedback, teachers need to have a good understanding of where students are, and where they are meant to be—and the more transparent they make this status for students, the more students can bridge the gap between their current understanding or performance to the success points. Chan et al. (2014) consider any instruction that is aligned with clear learning targets, and which yields observable student responses allowing teacher assessment of student progress, useful for providing effective feedback.

The present focus is on classroom participants' in-lesson verbal feedback, which constitutes the "ability to react to the learning during the learning" (Hattie 2019, p. 96). Wiliam (2006) argues for the use formative assessment as it allows feedback to be given during instruction. The main purpose of feedback, Hattie (2009) argues, is to reduce discrepancies between current understandings and performance and a learning intention or goal. Further, effective feedback answers three questions: "Where am I going?" (learning intentions/goals/success criteria), "How am I going?" (self-assessment and self-evaluation), and "Where to next?" (progression, new goals). The provision of verbal feedback in relation to learning intentions brings into focus the interactions among classroom participants.

8.3.2 Feedback and Classroom Interactions

The nature and participatory structure of classroom talk shape the opportunities for the teacher and students to give or seek feedback and broadly, ascertain the space for assessing student learning (Hussain 2022). Feedback becomes evident in the questions, comments or remarks surfaced by the teacher or students. Teachers persistently provide verbal feedback to indicate the adequacy or correctness of students' solicited or initiated statements (Storch and Wigglesworth 2010). Frequently, the teacher asks a question, student/s respond, and the teacher evaluates in a 'Ping-Pong' pattern (Good and Brophy 1987). Typically, such *Initiation-Response-Evaluation* (IRE) (Cazden 1988) interactions or presentational talk (Barnes 2008) constrain the space for formative feedback, which facilitates student understanding or performance. IRE or IRF (*Initiation-Response-Feedback*) interactional patterns are largely characterised by the teacher asking inauthentic questions in which students are expected to provide responses pre-determined by the teacher (Nystrand and Gamoran 1991). Contrastingly, exploratory talk permits working on one's understanding, enabling the speaker to reshape old knowledge in the light of new ways of 'seeing' things (Barnes 2008). Exploratory talk constitutes critical but constructive engagement with one other's ideas (Mercer 1995). Similarly, dialogic talk constitutes a form of understanding talk characterised by purposefulness, reciprocity and cumulation (Alexander 2004). Dialogic interactions are not commonly observed in classrooms (Nystrand et al. 2003) as school culture often imposes conversational 'ground rules' on classroom participants, which discourage students' reasoning, extended contributions, question posing and evaluation of peers' responses (Mercer and Howe 2012).

From a socio-constructivist viewpoint, feedback denotes shared and individual interpretations developed through dialogue, sense-making and co-construction among classroom participants (Price et al. 2011), facilitated through exploratory talk (Barnes 2008) as well as dialogic talk (Alexander 2004). The dominance of constructive, student 'dialogic' talk in a classroom characterises a 'formative assessment' culture (Clarke 2008). Dialogic feedback opens up space for knowledge exploration because of the collaborative or reciprocal association between students and teachers (Ajjawi and Boud 2017) as students act on the feedback to improve their work or learning strategies (Carless and Boud 2018). Teachers need to recognise students as co-constructors of knowledge and give them opportunities to engage with, interpret and use information over the course of classroom interactions, and these opportunities are contingent on teachers' pedagogical decision-making.

8.3.3 Teacher Pedagogical Reasoning and Formative Assessment

This article illustrates the pedagogical reasoning of primary science teachers as "active, thinking decision-makers who play a central role in shaping classroom events" (Borg 2006, p. 1) including the use of formative assessment. Pedagogical reasoning relates to teachers' pedagogical content knowledge (PCK)—a way of thinking that permits teachers to transform their subject matter knowledge into forms learners can understand (Shulman 1987). Teacher reasoning comprises actions that a teacher undergoes during the teaching process: comprehension of subject knowledge, transformation of subject knowledge into teachable representations, instruction, evaluation of student learning and teacher performance, reflection, and new comprehension. Directly relevant to the transformation stage, is teachers' PCK or *pedagogical reasoning*, which develops as teachers plan lessons, teach, adapt instruction to meet learners' needs, and reflect on teaching. PCK denotes instructional expertise—a combination of subject matter knowledge and the knowledge of how to teach it.

Teachers' actions are more accurate representations of their knowledge (Berk and Hiebert 2009), and the current focus is on tapping into teachers' pedagogical reasoning drawing on classroom observations and teacher interviews. Specifically, the focus is on teachers' enacted PCK (ePCK)—the unique subset of knowledge and skills that a science teacher draws on and plays out while planning, teaching and reflecting on a lesson (Carlson et al. 2019). Arguably, the present insights surface what Alonzo et al. (2019) refer to as teachers' plan-teach-reflect cycle, which largely ascertains formative assessment practices.

Teacher decision-making constitutes a key component of classroom assessment practices (Mandinach and Schildkamp 2021) but availing of assessment information to make teaching decisions is a complex process. Teachers make numerous decisions-in-action with little time to think (Korthagen 2010). Moreover, they may not have the capacity to translate information into insights (Datnow and Hubbard 2015). The use of AfL requires time and preparation (Datnow et al. 2021). An increased focus on using assessment for accountability (DeLuca et al. 2019) influences AfL practices as in Lee and Coniam's (2013) Hong Kong study in which teachers recognised the value of formative feedback but focused on detailed error corrections because of school policy and examination stress. Tay and Lam (2022) observed an error-focused feedback approach in five Singapore schools similar to the findings from Ha and Murray's (2021) Vietnamese study. Generally, change in teacher practice is difficult and takes time (DeLuca et al. 2019) and formative assessment may be especially difficult to enact because of the interactive and in-the-moment characteristics of AfL (Brown 2019). The enactment of teaching decisions—planned or unplanned—are captured via coding analyses of lesson videos in the present study.

8.4 The Study

The data is drawn from the Core Research Programme[1], a baseline study of pedagogical practices in Singapore. Based on stratified random sampling, data was collected in the school Terms 3 and 4 (2015) and Terms 2 and 3 (2016) from 10 primary and secondary schools. Pertinent to the present focus, the sample included five primary schools: three Government, one Government-aided and one Special Assistance Plan school[2] in line with the distribution of types of schools in Singapore. Each school nominated one science teacher with a minimum of four years teaching experience, teaching an academically 'average' performing class. Researchers observed a complete unit of school curriculum work i.e. a series of thematic lessons for each teacher participant. 49 lessons corresponding to five curricular units at the Primary 5 level (students' age range: 10–12 years) in five schools were observed (and video-recorded). Thematically, three units focused on 'Electrical system' (taught by T1, T2 and T3 respectively) and the other two related to the 'Human system' (taught by T4) and 'Plant system' (taught by T5). To illustrate the findings, each teacher's lessons are referred to as L1, L2 and so on.

Each classroom observation was followed by a brief teacher interview (audio-recorded), which typically included these questions: *What are the lesson objectives? How did you go about preparing for this lesson—materials/activities? Overall, how do you think the lesson progressed? How do you know lesson objectives were achieved? Were there any unexpected changes over the course of the lesson?* Content and thematic analyses of post-lesson interviews captured teachers' immediate sense of the lesson, reasoning about achievement of lesson objectives, and rationale for unanticipated shifts in their instructional approach, if any. Owing to logistical reasons, 40 post-lesson interviews were conducted. Data-driven, inductive analyses of an end-of-unit, semi-structured interview (audio-recorded) with each teacher elucidate teachers' pedagogical beliefs and reasoning.

Lessons were video-recorded using two high-definition video cameras—one captured the teacher (who wore a wireless microphone), and the second focused on the students. Additionally, two audio recorders supplemented the video recording. The researchers involved in the classroom observations subsequently engaged in coding analyses of lesson videos. Lesson videos were coded (in Microsoft Excel) in 5-minute intervals ('phases') based on a largely binary, domain-specific coding scheme, which was developed and refined based on the syllabus (MOE 2014), broader theoretical understandings, and insights from the data. Coding analyses recorded whether an instructional event happened (or not) during a 5-minute lesson phase. Coding indicators (not mutually exclusive) included student learning activities, classroom talk, inquiry processes, and visible learning practices. Inter-rater reliability was monitored, and validity of the codes was reinforced through intensive discussions among the coders and engagement with domain-specific experts. SPSS was used for compilation and statistical analyses of the coded data from which the descriptive data is drawn.

8.5 Findings

The findings from Kwek et al.'s (2018) large-scale study are presented according to the research questions. To address the initial research question about formative assessment practices, descriptive data as well as excerpts from the lessons and teacher interviews are presented. To answer Research Question 2, key themes from post-lesson interviews are surfaced pertaining to teachers' pre-lesson and in class teaching decisions.

8.5.1 Formative Assessment Practices

Formative assessment entails making the process of teaching and learning visible to both teachers and students (Hattie 2009). Table 8.1 shows that teachers communicated lesson objectives in more than two-fifth of the lessons. Teachers frequently conceptualised learning objectives in terms of students' learning goals, as evident in one teacher's lesson objectives (on the whiteboard): *To find out how the number of batteries affect the brightness of a bulb.* Teachers related concepts to familiar contexts in about three-fifth of the lessons. In one lesson, students worked in groups to light up a bulb, and one group noticed a spark in the bulb. The teacher pointed out that it was a short circuit, similar to what they may have experienced at home.

The findings show reasonable evidence of teachers activating students' prior knowledge by surfacing concepts or skills previously learnt (Table 8.1). T5's learning objective in L1 was "to connect them to what they have already learnt" about animal reproduction. To introduce the concept of conductors and insulators of electricity, T2 activated his students' prior knowledge, "Last year you learnt the topic on heat. Have you learnt this term 'conductor'? What do you know about it?" Teachers recapitulated the lesson content in slightly more than one-third of the lessons. T5's observations after L5 surface a pertinent issue: "I think it would have been better if I had more time at the end. Didn't really have enough time to round this off properly."

For the purposes of coding analyses, classroom participants' feedback was coded during students' engagement in scientific processes such as investigations or problem-solving activities, which provide space for students to acquire inquiry skills. Inquiry entails students actively thinking and engaging in procedures scientists follow to carry out investigations (Minner et al. 2010). In the observed lessons, students worked on practical tasks such as identifying materials that are conductors and non-conductors of electricity, or testing (using concrete manipulation) various ways of lighting up the bulb. The syllabus foregrounds opportunities for students to use concepts, and integrate scientific skills and processes for inquiring into phenomena around them (MOE 2014). Aligned with the essential features of inquiry (*Question, Evidence, Explanation, Connections* and *Communication*), the coding indicators for inquiry were drawn from Bybee's (1997) '5E' model: *Engage, Explore, Explain, Elaborate,* and *Evaluate.* The normative codes offer a realistic appraisal of the extent

Table 8.1 Primary 5 science (2015/2016): visible learning practices

Visible learning practices (rank ordered)	P5 (N = 49)
Teacher relates concept/topic to everyday life	60.5%
Teacher checks students' pre-requisite concepts and/or skills	37.2%
Teacher tells students what they have learnt previously	44.2%
Teacher communicates learning goals and outcomes	41.9%
Teacher recapitulates and summarises lesson content	37.2%

and manner in which the lesson gives evidence of inquiry via investigative activities and/or problem-solving scenarios. Classroom participants' feedback was coded when one or more of these '5E' features became evident thus, surfacing evidence of the role of feedback in enabling disciplinary practices such as posing and responding to questions, designing investigations, and communicating understandings (MOE 2014).

Table 8.2 shows that teachers provided almost twice as much detailed feedback than evaluative feedback during scientific investigations (see Appendix 1). Arguably, teachers focused on facilitating student understanding or performance, as evident from one teacher's detailed feedback when a student noticed a spark while working on circuits:

> Yeah that means there's current and the current has gone into umm, that means you're not connecting it properly. You're connecting it wrongly. If it's very hot and you see spark, it means your connection is wrong. It's a short circuit. Later on you'll learn about short circuit.

In a lesson on seed dispersal, students conducted an experiment to ascertain the effect of wing-like structures on the flight time of seeds. Reinforcing domain-specific criteria, the teacher offered detailed feedback to a student whose readings varied considerably:

> When you have readings like that, you are not sure which is the most reliable reading because they are not consistent. You have 3, you have 4, then suddenly a 6.8 which is quite far away from the other two readings, almost a 3-seconds difference, so under this circumstances, I will take a fourth reading and see which the fourth reading is nearer to 3 or 6.8? If my fourth reading is 4 point something, I will ignore the 6.8. Here repeating an experiment allows me to have more reliable readings. If all three of my readings are about the same, I can be quite sure that my experiment is reliable.

Students too provided feedback but far more to their teacher than peers, as Table 8.2 shows (see Appendix 1). A student offered feedback on her classmate's performance during a group investigation, "No, the light bulb is supposed to be here." Admittedly, the general noise in the classroom, limitations of the audio and video recording setup, and missing contextual information made it difficult to capture instances of student-to-student feedback.

Other than the use of feedback and visible learning practices, the space for formative assessment of student learning relates to the structure and nature of classroom interactions and activities. In terms of participatory structures, students had few

Table 8.2 Primary 5 science (2015/2016): feedback during scientific investigation

Feedback during scientific investigation/ problem-solving	P5 (N = 124) Fraction of phases per lesson
Teacher evaluative feedback	0.29
Teacher detailed feedback	0.56
Student to teacher feedback	0.42
Student to student feedback	0.27

opportunities to collaborate in pairs or groups (Table 8.3). For his initial lesson, T2 deliberately incorporated a hands-on session into the learning goals: “To allow students to interact with peers to understand the components (electrical) better.” In about one-third of the lesson phases, students reviewed or revised their performance.

In more than two-third of the lesson phases, students listened to the teacher’s exposition (Table 8.3). Similarly, IRE interactions (Cazden 1988) are dominant, which is understandable given that teachers mainly asked closed questions (Table 8.4). Referring to a circuit diagram (on the whiteboard), a teacher asked, “Is this an open or closed circuit?” Students offered one of the two responses based on the visual image. Contrastingly, approximately a quarter of the lesson phases reveal teachers’ open-ended questions, which potentially elicit multiple or divergent responses from students (Table 8.4). While working on connecting wires to a bulb and a battery, a student exclaimed, “We found another way (to light up the bulb)!” The teacher followed up, “Good, any other way?” Beyond mere evaluation of student performance, the teacher’s open question in this instance helped the student consider alternatives. Time constraints often thwarted teachers’ attempts to explore issues. For instance, T1 “wanted to go a bit faster, even go through the open-ended questions in their workbook” in L7 but postponed those to the next lesson.

The predominance of IRE interactions (Cazden 1988) and the negligible presence of whole class discussions (Table 8.3) suggest that the three-turn sequences were largely those in which the third turn performed an evaluation (‘E’) of student understanding or performance. Nassaji and Wells (2000) identify six functional moves that the third turn displays: evaluation, justifications, counterargument, clarification, meta-talk and action. While the analyses in the present study does not capture the functional moves of the third turn in three-turn interactions, the coding instrument provides evidence of different kinds of classroom talk, as Table 8.5 shows (see

Table 8.3 Primary 5 science (2015/2016): student learning activities

Learning activities (rank ordered)	P5 (N = 569) Fraction of phases per lesson
Listening to the teacher’s exposition	0.69
Participating in initiation-response-evaluation (IRE) sequences	0.69
Checking/reviewing/revising	0.34
Pair/group work	0.16
Practice (e.g., on worksheet/s, notebook, computer)	0.10
Participating in whole class discussion	0.06

Table 8.4 Primary 5 science (2015/2016): teacher question type

Teacher question type	P5 (N = 347) Fraction of phases per lesson
Closed question	0.88
Open question	0.27

Appendix 2). Explanatory talk refers to classroom participants providing reasons or explanations for their initial statement, which denotes the clarification function of the third turn. Similarly, justification talk relates to the functional move of justification. Broadly, the nature of classroom talk offers insights into how the third turn in IRE interactions potentially facilitates formative assessment of student learning.

In terms of classroom talk, more than two-fifth of the lesson phases, explanatory talk becomes evident (Table 8.5) indicating that feedback often entailed seeking clarification followed by explanation (see Appendix 4). IRE interactions in the observed lessons barely reveal the functional move of justification (Nassaji and Wells 2000). Justification talk, which potentially enables students' domain-specific understandings is practically absent (see Appendix 5). Procedural talk shows reasonable evidence, which implies that scientific skills and processes are emphasised but students hardly engage in domain-specific justification for the use of procedures (Table 8.5). Based on procedural talk, T5 gauged that her students "seem to have different variables that they are investigating" (L7). Factual talk, which entails students acquiring or retrieving information is dominant. T5 assessed that L3 progressed well given that students "know their facts" and in one lesson, she asked her students to list the reproductive parts of a female flower.

Classroom talk that permits students to make conceptual connections surfaces evidence of student learning. Teachers made students' learning visible by linking concepts to daily life contexts (Table 8.1) thus, facilitating conceptual linkage with the thematic focus (Table 8.5). Additionally, teachers provided students opportunities to forge conceptual links via explanatory talk. Equipped with his diagram, a student explained how the magic light bulb worked:

> So in the light bulb, this area, may have free zone electrons inside. So when a metal object touches it, the electrons are transferred to the metal object then transferred back, probably touching a circuit that powers the light bulb inside.

Temporal connections show a weak presence, which may be cautiously attributed to teachers' pre-lesson planning (Table 8.5). Notably, only T4 highlighted that she "looked through student work" in the previous lesson/s "to diagnose misconceptions" that need to be addressed.

The above findings pertaining to students' learning activities (Table 8.3), the structure of classroom interactions (Tables 8.3 and 8.4) and the nature of classroom talk

Table 8.5 Primary 5 science (2015/2016): classroom talk

Classroom talk (rank ordered)	P5 (N = 569) Fraction of phases per lesson
Factual talk	0.58
Procedural talk	0.38
Explanatory talk	0.42
Conceptual connections	0.29
Temporal connections	0.07
Justification talk	0.01

(Table 8.5) portray the opportunities for formative assessment of student learning in the observed lessons. Classroom talk mediates knowledge and cognition (Atwood et al. 2010), and its structure determines the space for student participation (Freebody 2004). Tasks comprising learning activities form the backbone of classroom interactions and frame how students think and learn (Blumenfeld et al. 1987) ascertaining the quality of knowledge work students engage in (Hogan et al. 2013). Thus, a focus on classroom talk and student learning activities merits attention to the knowledge focus, captured by a range of coding indicators in the present study (see Appendix 3).

Table 8.6 indicates a dominance of factual knowledge in line with the substantial evidence of factual talk (Table 8.5). In one lesson, students were asked "What have you learnt about the bulb?" Leveraging on his factual knowledge, one student stated, "The bulb has a filament, a glass cover, a metal casing and a metal tip." Holding up the bulb, he pointed out, "The metal casing is here. The metal tip is this one here." Procedural talk and procedural knowledge show similar trends (Tables 8.5 and 8.6). T5 targeted procedural skills in L8: "… to test out, to confirm, to conduct the experiment, to confirm that the presence of the wing really makes a difference" as students engaged in scientific investigation.

Conceptual knowledge is evident in about half the lesson phases as classroom talk is often explanatory in nature, and reasonably focuses on conceptual connections (Tables 8.5 and 8.6). In their interviews, most teachers highlighted the importance of concept development. Teachers' focus on enabling students' conceptual understanding is reflected in T2's assertion after L5: "Wanted to introduce conductors and insulators but I didn't use the term yet. I want them to understand the concept, make some meaning out of the terms, the terms are just there to label them." To reinforce his students' understanding of close/open circuits (prior to working with the battery and bulb), T2 "changed the order of the activities" thus, ensuring it is not a "very mechanical thing."

Returning to Coffey et al.'s (2011) argument about discipline-specific formative assessment, the negligible presence of epistemic knowledge (Table 8.6) and justification talk (Table 8.5) is noteworthy. Epistemic knowledge denotes knowledge frameworks pertaining to the rules and criteria used to develop and/or judge what counts as scientific (Duschl 2003). Reflecting on her students' investigation on fruit dispersal, T5 noted, "… the scientific part not too sure but I think the engagement level is there," indicating her focus on students' disciplinary understandings. T1 had

Table 8.6 Primary 5 science (2015/2016): knowledge focus

Knowledge focus (rank ordered)	P5 (N = 569) Fraction of phases per lesson
Factual knowledge	0.59
Procedural knowledge	0.39
Conceptual knowledge	0.49
Epistemic knowledge	0.04
Metacognitive knowledge	0.04

a disciplinary rationale for introducing graphite (non-metal) as a good conductor of electricity (L6):

> For me, graphite is not in the syllabus but it's good to know that most of the time generally metals are good conductors but there are other components. So I don't want them to think that science is like always fixed - there are exceptions to it.

Metacognitive knowledge is hardly emphasised (Table 8.6), which suggests that classroom interactions and learning activities barely permitted students to reflect on their learning or share their reflection with peers. In their interviews, teachers hardly mentioned students' reflection on their learning. Collaborative work shows modest evidence (Table 8.3), and evidentially, students seldom engaged in peer assessment or self-assessment.

In view of the initial research question, this section portrays the nature of AfL practices in Primary 5 science classrooms. Teachers articulated learning goals, activated students' prior knowledge, and forged conceptual connections. Teachers' feedback focused on informing students' learning as they engaged in scientific investigations. Students were seldom activated as learning resources for themselves or one another. Arguably, the dominance of IRE interactions (Cazden 1988) and teachers' closed questions constrains the space to formatively assess student learning. The negligible presence of justification talk and epistemic knowledge possibly limits discipline-specific formative assessment (Coffey et al. 2011). Formative assessment practices relate to teachers' pedagogical decision-making, as elucidated below.

8.5.2 Teacher Pedagogical Reasoning: Formative Assessment

The post-lesson interviews reveal that teachers conceived lessons from the perspective of students' learning goals. For his initial lesson, T2 intended "for students to understand the components (electrical)" and therefore, included a hands-on session. T1 reasoned about the need "... to transform their knowledge—concepts to be able to answer exam-type questions." Infrequently, teachers focused on the instructional activity in lesson planning. For instance, T5 intended "to run through the topical worksheet with them" in her final lesson. Teachers often articulated lesson objectives in learning-oriented terms such as *know, identify, understand, reinforce* etc.

Given their focus on students' learning, teachers attempted to elicit verbal responses and formatively monitor student work. T2 observed after L3: "Through the questions, I could see most understood the concept of closed, open (circuits), very sure handful who still don't get the concept." T1 was convinced that the lesson objectives for L3 had been achieved: "About 70% were able to grasp—hopefully by next week 100% will be able to integrate the symbols. When I go around, I saw two groups were not sure, a bit slower." After L2, T5 noted that students were able to identify parts of a flower, and one student "was able to surface after some probing." Monitoring student work helped T2 gauge her students' learning needs (L6): "When

I went to different groups, I feel I met the objectives. They understand parallel/series circuits but have difficulty with matrix sets."

While teachers largely assessed students' understanding based on explicit or implicit evidence in the classroom, they also assessed student work after the lesson and occasionally, tweaked the focus of subsequent lesson/s. T2 noted after L1: "Most students were able to answer but I want to see worksheet then I will know whether objectives have been met or not." T4 mentioned the need to see her students' "actual work" after L3. After L5, T2 felt that the learning objectives had been met based on his formative monitoring but intended to employ "informal testing in the next lesson" to gauge his students' understanding of circuits. Ascertaining that the learning objectives for L4 had been partially met, T1 decided that she would ask students to articulate their ideas in the subsequent lesson.

In terms of lesson preparation, teachers largely drew on MOE resources such as the 'Teaching and Learning Guide' (TLG), the syllabus, textbook, workbook etc. as well as school-based materials. T4 mentioned sourcing for additional materials online mainly on YouTube and via Google search. Infrequently, teachers reported examination of student work as the bases for their lesson plan. T2 planned L7 based on his marking, which revealed that "... the switch was clearly a misconception—one of the students thought the circuit must have a switch, so they wrote there is no switch, so the bulb does not light up." For planning lessons, T4 "looked through student work to diagnose misconceptions" from previous lesson/s.

Teachers' immediate sense of the lesson primarily related to time constraints and how it impacted their students' conceptual grasp, which often resulted in modifications to their lesson plan. Reflecting on L3, T2 noted, Slower than expected—had "planned out quite a bit for today's lesson but had to cut. This class is fast on graphical information but slow in writing." Similarly, T4 pointed out that she was "rushing through so a lot of them did not catch the verbal instruction" in L9. Referring to L11, T5 pointed out, "If I had two periods and I don't have to rush, probably I would have spent more time there (transition of flower into fruit), like really allowing them to explore." T2 noted: "It's not about content—content five periods can be done to explain the concept. But when it comes to answering it is a whole different thing."

Students' queries and responses as well as affective and logistical aspects often resulted in teachers altering their pre-lesson pedagogical decisions. T5 noted the "hoo-ha" in L2 when students brought up the differences between an insect and an animal. Similarly, her students raised unanticipated questions about flowers being in bunches and the colour of the seed in L4 and L5 respectively, which made the teacher change her lesson plan. Teachers reported students being quiet, bored or excited during the lesson. T2 mentioned resolving his students' dispute about sharing equipment, as one student cried during the lesson (L2). T5 noted the loss of time because of the difficulty in finding a suitable venue for students' investigation (L8). T1 "did not expect apparatus not to work" (L6) and similarly, T2 observed that "lots of sets had a faulty bulb so that became a challenge" (L6).

To address the second research question, teachers were more inclined to teach towards learning objectives than teach to the instructional activity. Teachers gauged their students' ongoing learning via formative monitoring, and elicitation of verbal responses both in whole class and group work settings. Frequently, time constraints caused teachers to alter their pre-lesson teaching decisions. Unexpected student initiations as well as affective and logistical issues caused teachers to modify their lesson plan. These insights into teachers' pedagogical decisions and how they manifest into formative assessment practices are now discussed.

8.6 Discussion

The present findings suggest that Primary 5 science teachers in Singapore classrooms employed formative assessment practices to inform student learning and their own instructional repertoires, leveraging on the "ability to react to the learning during the learning" (Hattie 2019, p. 96). In the local context, the study surfaces the challenges of realising the curricular focus on AfL and students' self-directed learning in line with inquiry-based learning (MOE 2014).

The observed lessons show considerable evidence of the teacher's exposition and IRE interactions (Cazden 1988), which is pertinent given that the "power of feedback is rarely operationalised during these soliloquies" (Hattie 2012, p. 138). Teachers have a propensity to engage in transmission methods of teaching rather than actively involving students in their own learning (Gillies 2012). Similarly, Zhai and Tan (2015) found that teachers in three elementary science classrooms in Singapore prioritised whole class instruction and a teacher-directed learning environment. Teachers tend to rely on the routines they have developed (Eraut 2004) including their orchestration of classroom talk.

Questioning is a significant part of teaching and science talk (Chin 2006) and teachers predominantly asked closed questions (Table 8.4). The findings are noteworthy given that the quality of teacher questioning and the follow up on responses play a role in drawing out student thinking (Black and Harrison 2001). Teacher questioning is especially important in an inquiry classroom where teachers need to purposefully craft questions, which enable students' engagement with cognitive processes such as explanation, prediction, interpretation, inference, reflection and so on (Chin 2006). Teachers asked open-ended questions in more than a quarter of the lesson phases (Table 8.4). van Zee et al. (2001) observed that teachers elicited student thinking by asking questions that developed conceptual understanding. Eliciting information about students' understanding via questioning constitutes a key formative assessment tool (Chin 2006), and teachers in the study explicitly focused on eliciting verbal responses to surface evidence of students' conceptual understanding. The present findings also illustrate that time constraints hinder teachers' ability to engage students in discussions via open-ended questions especially so as students in Singapore classrooms do not readily articulate their ideas, or volunteer unsolicited information (Chin 2006).

While teachers' closed questions and IRE sequences (Cazden 1988) are dominant, the findings suggest some evidence of IRF interactions: initiation, response, and follow-up or feedback (Sinclair and Coulthard 1975)—the third move may not necessarily be an explicit evaluation of the student's response. Specifically, the data surfaces instances in which the third turn performs functions such as clarification and justification (Nassaji and Wells 2000). The evidence of explanatory talk (Table 8.5) suggests that the third turn often performed the function of engaging students via invitations, requests for clarification, elaboration, follow-up questions etc. Based on several Year 7 science lessons, Chin (2006) found that the F-move in IRF exchanges was often supportive in that it embedded a further question that stimulated deeper thinking. Teachers in Chin's study built on students' accurate responses to enhance their conceptual understanding, or challenged students with further question/s in case of incorrect response/s. Triadic dialogue sequences (Lemke 1990) can potentially scaffold students' extension of knowledge through further supportive dialogue (Bruner 1986; Vygotsky 1978) but teachers' tendencies towards closed questions possibly stifles productive thought (Table 8.4). Also, the weak presence of whole class discussions (Table 8.3) suggests that in very few instances, the IRE or IRF structure expanded into an IRFRF discourse pattern that supports dialogic interaction (Mortimer and Scott 2003) thus, limiting exploration of students' ideas and understandings. On a cautionary note, the third turn in three-turn sequences represents a contingent and interpretative undertaking by the teacher, which may not be easily captured by pre-determined functional categories (Lee 2007).

Other than their classroom enactment, the study elucidates teachers' beliefs about educational processes, which contribute to their pedagogical actions (Rubie-Davies et al. 2012). Teachers believed it was their duty to help students excel in high-stakes examinations, and considered it important to teach examination techniques, writing answers succinctly, and using scientific language. Beyond mere evaluation of student work, teachers attempted to leverage on the facilitative rather than the directive function of feedback (Black and Wiliam 1998) as their strong preference for detailed feedback indicates (Table 8.2). They employed feedback to scaffold students' learning (Galton 2007), as a cognitive support mechanism (Shute 2008). The findings are unsurprising given teachers' preference for extensive error-focused feedback in Singapore's exam-oriented classrooms (To et al. 2023) where they are expected to provide corrective feedback to improve students' examination results (Ratnam-Lim and Tan 2015). The negligible presence of justification talk (Table 8.5) and epistemic knowledge (Table 8.6) indicates that infrequently, teachers made direct references to the knowledge and nature of the discipline. Inquiry entails opportunities for students to grasp the methods by which knowledge is developed and validated in a discipline (Mansilla 2005) thus, teacher participants' use of feedback as formative assessment was not necessarily grounded in disciplinary understandings (Coffey et al. 2011).

Availing of student feedback (Table 8.2), teachers often exercised pedagogical agility and changed their lesson plan to address students' queries or misconceptions. Earle (2021) found that some primary science teachers made changes within the lesson while others provided follow-up activities or made long-term adaptation to teaching practices. On the contrary, teachers in Haug and Odegaard's (2015) study hardly used their increased access to student thinking to promote learning or revise their instructional decisions. Boney (2014) noted that teachers' decision-making operated in two modes: in intentional planning for instruction, and spontaneously during instruction. Tasks as set up by the teacher may differ from those implemented (Stein et al. 1996) i.e. *tasks as workplans* are effectively different from *tasks in progress* (Seedhouse 2005). Besides student initiations, teachers often modified their instruction because of inadequate time. Seung et al. (2012) observe how the dilemmas teachers face in terms of what they conceptualise, plan, and do in class hinder the development of PCK. Science teachers' PCK is often tacit and challenging to articulate, capture and portray (Loughran et al. 2012) typically because of time constraints, which often cause teachers—including those in this study—to alter their pedagogical decisions during lessons.

Teacher participants gathered, interpreted, and acted on information about students' learning (Bell and Cowie 2001) via explicit verbal responses and implicitly, through formative monitoring of student work including non-verbal cues. Arguably, teachers considered assessment as embedded in classroom instruction (Popham 2008). However, based on a sample of 39 primary science teachers in Singapore, Tan and Tan (2011) found that some teachers view assessment as separate from instruction. In the present study, teachers' informal assessment practices in the form of ESRU cycles become evident: the teacher *elicits* a question; the *student* responds; the teacher *recognises* the student's response; and then *uses* the information to support student learning (Ruiz-Primo and Furtak 2007). Evidentially, these ESRU cycles fuelled teachers' in class decision-making as well as their lesson planning. Also, teachers framed lesson objectives in terms of students' learning goals, which contradicts early research suggesting that teachers teach to the instructional activity and treat the activity as the basic unit of planning (Clark and Yinger 1979; Shavelson et al. 1977). Overall, the insights into teachers' pedagogical reasoning and their classroom enactment offer some key takeaways.

8.7 Implications and Conclusion

The present findings surface key pedagogical implications for classroom practice:

1. The dominance of teacher-talk and IRE in the study foreground a fuller picture of feedback and uptake in line with a learner-centred feedback approach (To 2022). Student–teacher dialogues offer space for teachers to respond and reorient students' thinking (Black and Wiliam 1998). Besides giving and seeking feedback, taking up feedback from students can meaningfully inform students' ongoing learning, and provide useful bases for teachers to adapt and adjust their instructional repertoires.
2. Given the implications of teachers' closed questions for AfL practices, emphasising inquiry-based strategies such as questioning (and follow up of student responses) can help teachers better leverage on the potential of in-lesson verbal feedback. The kind of questions teachers pose ascertain the nature and quality of the student thinking (Black and Wiliam 1998). Teachers' use of questions

and how feedback potentially supports learning and shapes students' thinking (Hardman 2008) need more attention.

3. The negligible presence of metacognitive knowledge warrants a stronger focus on students' metacognitive learning via reflection, self-assessment, reflexive talk, and specific metacognition strategies entailing both knowledge and regulation of cognition (Flavell 1979). Additionally, encouraging collaborative work can facilitate more deep and meaningful learning (Garrison and Akyol 2015) and permit sharing of reflection, peer feedback and peer assessment. Consolidating student learning and recapitulating lesson content also need to be strengthened—"a robust lesson will fall short of its full potential if the lesson doesn't include a solid closure" (Fisher et al. 2017, p. 71).

Related to the above recommendations for practice is the underlying issue of time constraints, which needs to be systematically addressed at the policy level. Specifically, lesson duration and allocation of lessons in the school timetable need to be more conducive for discussion, collaborative learning, reflection tasks, lesson consolidation etc. The logistical constraints teachers experience also require more attention. The findings inform local professional development initiatives such as the SkillsFuture for Educators (MOE 2020), which emphasises learner-centred assessment and students' self-directed learning. Broadly, the study surfaces valuable insights for teacher training:

1. The findings surface how AfL practices hinge on teaching decisions thus, enabling teachers to be more explicit about their tacit decision-making can support them in more formative use of assessment information to enable student learning (Earle 2021). Teachers' tacit processes and 'gut instincts' need to be made more explicit to explore how and to what extent formative processes can be incorporated into classroom practices (Buck and Trauth-Nare 2009). More space for teachers to (re)consider the implications of their intentional lesson planning and in situ decision-making can help them better grasp how AfL is employed in their specific classroom context.
2. Beliefs about enabling students' academic performance shaped teachers' feedback practices, which suggests the need for a stronger focus on teachers' assessment literacy—a developmental process that requires teacher's reflection and critical evaluation of their diverse use of assessment (DeLuca et al. 2016). Teachers' assessment literacy is instrumental to AfL as their capacity to enact feedback is complex and contextual, involving their assessment identity, beliefs, dispositions towards classroom enactment and perceived role as assessor (Looney et al. 2018).
3. The study warrants a greater focus on anchoring formative assessment practices in disciplinary ideas and reasoning in line with inquiry-based learning. Assessment literacy or capability also necessitates an understanding of the subject being taught thus, teachers' professional learning around assessment needs subject-specific elements to render it usable in the classroom (Earle 2021). Discipline-specific formative assessment potentially facilitates justification talk and epistemic knowledge thus, facilitating knowledge construction in the inquiry classroom.

The present findings advance our theoretical understanding of science teachers' pedagogical decision-making based on empirical evidence of formative assessment practices. Insights into teachers' PCK from varying school and classroom contexts offer a nuanced picture of teachers' AfL practices. Specifically, the findings elucidate how the use of feedback, visibility of teaching and learning, classroom participatory structures and nature of classroom interactions ascertain the space for formative assessment of student learning. By tapping into teachers' pedagogical decision-making via classroom observations and interviews, the findings depict the "pervasive tension between thinking and doing that characterises the work of teaching" (Anderson and Peck 2007, p. 19). Overall, the article showcases how the curricular emphasis on formative assessment and inquiry plays out in the Singapore classroom given teachers' lesson planning considerations and the ground realities of employing AfL in day-to-day teaching.

8.8 Note to Our Future Colleagues

As the technology entrepreneur, Elon Musk predicts, human language may be rendered obsolete in the near future given the significant move towards telepathic communication (Embury-Dennis 2020). Will directly linking human minds to computers eliminate the need for a 'language' to communicate in the classroom? Will verbal feedback to enable student learning become superfluous? I'd argue otherwise—while technology continues to make huge inroads into conventional modes of teaching and learning, the role of the teacher as the human interface for equipping students with desired knowledge, skills and dispositions remains undisputed. In the same vein, the modality of talk will remain critical to student learning. Unsurprisingly, teacher participants when asked (after the lesson) whether the learning objectives had been achieved (or not), frequently based their reasoning on students' verbal responses: "Because when I ask students questions, most of them are able to answer." One teacher participant noted, "Expressing an answer is a problem but that one really takes time." The investment of time and effort in enabling students' verbal expression may well be worth it as the nuances of the spoken word and the richness of interpersonal interactions are likely to remain germane to formative assessment practices.

At the time of writing this chapter, preliminary findings from an ongoing baseline study suggest Primary 5 Science teachers' frequent use of technology-embedded formative assessment in Singapore classrooms (Kwek 2023). Arguably, even as teachers increasingly leverage on technology to make students' thought processes visible, rapid strides towards telepathic communication are potentially disrupting the bases of making teaching and learning visible to both teachers and students (Hattie 2009). As new frontiers in technology inevitably shape classroom pedagogy, giving or seeking feedback via brain signals may not be uncommon in the 2050 classroom. Verbalising feedback will possibly be deemed a highly complex and valued skill for which students and even teachers need to be specially trained and encouraged.

Perhaps, classroom participants three decades later will be confronted with more complex decision-making in terms of their formative assessment practices: Should I just use a signal to convey this simple error correction to the class?; May be I actually have to talk to my teacher for improving on this.; To give him a fair chance to clarify the criteria, I'd better use words.; My teacher promptly reads my mind and helps me, so why ask? Endorsing Black and Wiliam's (1998, p. 16) key observations that the interaction between teachers and students during formative assessment lies at the "heart of pedagogy", we may increasingly need to grapple with the issue of defining the nature of the interaction. Unpacking the when, how and why of classroom participants' verbal responses towards formative assessment of student learning may perhaps be the most essential skill in the pedagogical toolbox of a teacher in the future. Eventually, it may boil down to (re)imagining the teacher, talk and technology to optimise student learning.

8.9 Notes

1. The 'CORE Research Programme' comprises CORE1 (2004–2007), CORE 2 (2009–2014), CORE 3 (2015–2023) including the OER 17/14 KBK project from which the present data is drawn, and CORE 4 (2023-present). With the chief objective of providing systematic, robust and purposeful descriptions of actual, everyday classroom pedagogy, CORE has conducted large-scale investigations into the nature of teaching and learning in Singapore schools since 2004. The data for this CORE Project no. OER 17/14 KBK has not been published previously but has been shared and presented with the Ministry of Education (Singapore) and other stakeholders as well as at key international conferences. The primary focus is on fast-cycle, timely reporting of findings to the ministry. CORE's well-informed, timely and evidence-based baseline descriptions on the state of pedagogical practices in Singapore schools impact policy reforms and programmes, curricular reviews, and classroom innovations.
2. Based on the regulating body of the school and the degree of financial assistance received, primary and secondary schools in Singapore are categorised as Government, Government-aided, or Independent schools. The curriculum, study materials, examinations, fee structure etc. in Government schools are fully regulated by the Government. Government-aided schools have some degree of autonomy but function according to Government regulations. Comparatively, Independent schools have greater autonomy. Special Assistance Plan (SAP) denotes a programme to cater to students who have high academic proficiency in their Mother Tongue and currently, the SAP focus is on the Chinese Language.

Acknowledgements This article makes use of data from the Core Research Programme: Baseline investigation of science pedagogy. The study was funded by the Education Research Funding Programme, National Institute of Education (NIE), Nanyang Technological University, Singapore:

Project no. OER 17/14 KBK. The views expressed in this article are the authors' and do not necessarily represent the views of NIE.

Appendix 1

Feedback type (Kwek et al. 2018)	
Coding indicator	Description
Teacher evaluative feedback	The teachers' verbal remarks that reflect the adequacy or correctness of the student's understanding or performance. Generally, it does not include any explanation, detail or guidance for future learning
Teacher detailed feedback	The teacher's constructive comments that draw attention to errors in student understanding or performance. The feedback enables students' modification of current performance and/or improved understanding. Based on Hattie and Timperley (2007), detailed feedback encompasses feedback at the task, process and/or self-regulation levels
Student to teacher feedback	Based on understanding/performance, the student gives or seeks feedback from the teacher. The feedback may be student-initiated or based on the teacher's invitation, and may be in any form e.g. question, clarification etc.
Student to student feedback	Student gives or seeks feedback from peer/s in the form of a question, clarification, confirmation etc. Feedback may be initiated by the students themselves say, during collaborative work or teacher-led contexts, or by the teacher as part of the lesson design e.g. gallery walk, pair and share

Appendix 2

Talk type (Kwek et al. 2018)	
Coding indicator	Description
Factual talk	Talk about propositional knowledge (facts, equations, definitions, algorithms etc.); factual talk often involves description
Procedural talk	Talk about *how* to complete a process or task; it can refer to general procedural issues such as inquiry methods, genres of work or more narrowly, task-specific scripts, rules, procedures, strategies, algorithms, heuristics or resources
Explanatory talk	Teacher/students give reasons or explanations for the initial statement; the reasons may be in the form of description, giving evidence, providing justification or making connections
Conceptual connections	Teacher/students make connections or comparisons within or across concepts, representations, examples/analogies, arguments and contexts

(continued)

(continued)

Talk type (Kwek et al. 2018)	
Coding indicator	Description
Temporal connections	Beyond a superficial reference, teacher/students make connections to content taught earlier in the lesson, or link the current lesson to previous lesson/s or subsequent ones, which enables understanding the current lesson content
Justification talk	Teacher/students identify and discuss the domain-specific epistemic norms (criteria and standards) used to establish the truth value, rigor, validity, reliability, authenticity, reasonableness etc. of a knowledge claim

Appendix 3

Knowledge type (Kwek et al. 2018)	
Coding indicator	Description
Factual knowledge	Knowledge of facts, definitions, terms, details
Procedural knowledge	Knowledge of procedures, algorithms, steps, skills; *how* to do the given task
Conceptual knowledge	Knowledge of the meaning of concepts/ideas, and relationships across facts/concepts/ideas within a larger structure
Metacognitive knowledge	Knowledge of cognition; knowing when or why to use strategies to enable learning; learning to learn
Epistemic knowledge	Knowledge of the nature of the discipline, and disciplinary standards to establish and validate knowledge

Appendix 4

The following extract shows the teacher's use of clarifying questions and explanations to draw comparisons (Fig. 8.1).

Extract.

1. Teacher: Ok so basically these are the circuits—they have two set-ups for you. Let me ask you—is there another way to study?
2. Student: Number of wires.
3. Teacher: Eh?
4. Student: Number of batteries.
5. Teacher: Number of batteries? Yes? Number of batteries how? They are all four batteries, two batteries?
6. Student: (Inaudible)

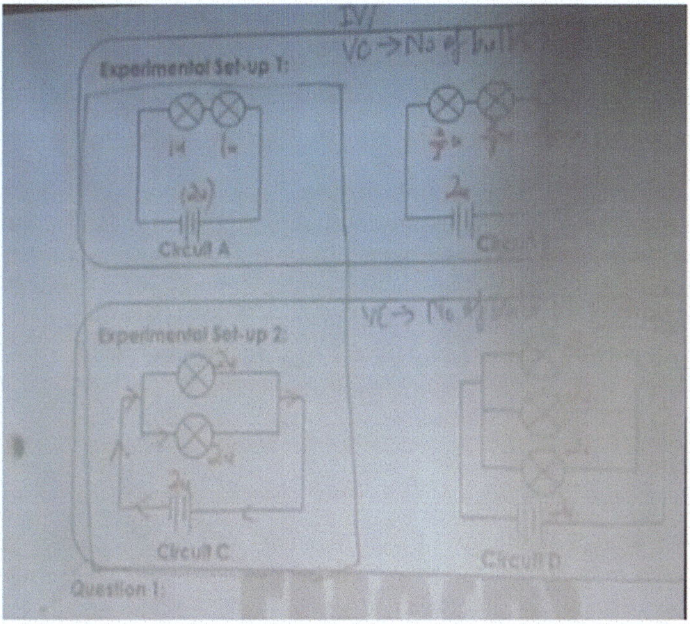

Fig. 8.1 Experimental set-up: comparison of parallel and series circuits

7. Teacher: When I, when I look at this everybody, when I look across like this (Teacher draws a line enclosing Circuit A and Circuit C), what am I trying to find out? What am I trying to find out if I am comparing Circuit A and Circuit C?
8. Student: (Inaudible)
9. Student: Brightness!
10. Student: It is brighter?
11. Teacher: Yes, but what am I comparing, what am I trying to find out?
12. Student: Whether splitting the…
13. Student: No, no whether…
14. Teacher: Go back to the variable change. What is the variable change?
15. Student: Number of bulbs.
16. Teacher: Is it the number of bulbs? Circuit A and Circuit C.
17. Student: Arrangement!
18. Teacher: Very good. So when I compare A and C, I am trying to ascertain if the arrangement has an effect ok? And if we do a mathematical calculation, we will know that it has an effect. Why? Because this (Circuit C) is a parallel circuit. Parallel circuits are selfish. So what do they do? They don't?
19. Student: (in chorus) Share.
20. Teacher: Ok they are not only selfish, they are also greedy. So they want a full share of the current. So you look over here—this is 2U alright (T points to the batteries in Circuit C) so they are going to take everything the batteries can give.

So over here 2U will go to this and 2U will go to this (T points to each of the two bulbs in Circuit C). Can you see that?

In the above extract, T4 attempts to clarify the student's reponse: 'number of batteries' (Turn 5). To help students grasp the rationale for comparing Circuit A and Circuit C, T4 probes deeper (Turn 7 and Turn 11). The teacher's explanatory talk (Turn 18) commences with some evaluative feedback ('Very good') after which she focuses on the arrangement of the given circuits and then invites her students to consider the 'selfish' characteristic of parallel circuits. The teacher further explains that the bulbs in parallel circuits obtain the full share of the current (Turn 20). Subsequently, T4 reinforces her students' understanding by explaining that the brightness of the bulbs in parallel circuits is unaffected by the number of bulbs.

Appendix 5

The following extract highlights T1's justification based on disciplinary criteria, as she reviews the work assigned in relation to the given experimental set-up: *Identify three variables, which have not been changed.*

Extract.

1. Teacher: Now three variables (T points to the question). There are some of you who gave me only one variable or two variables. Next, there is one person who just gave me 'bulbs' and 'switches.' What about the bulbs and switches? Is it the colour? They must be of the same colour.
2. Student: Number!
3. Teacher: So why are you writing 'bulbs and switches?' You are telling me just bulbs and switches. You are telling me they should be of the same colour. Same design is it? Same pattern? For science, you cannot just give … in science, you have to be very specific. Because if you just give 'bulbs and switches,' there are many things. What about the switch? Is it the same colour, same size, same what, same pattern, same design? You are talking about the number. And you need to write 'number.' If you just say 'switch,' we cannot give you the mark because what about the switch? Honestly, the colour of the switch eh it doesn't matter. Does it really matter?
4. Student: No.
5. Teacher: Does it, does it affect your experiment? No, it doesn't. What affects your experiment? Honestly, the type and the number will affect.
6. Student: Teacher, why will the number of switches affect?
7. Teacher: Because the switches act as resistance. It is something like it is more difficult—when you add all this, it's more difficult for the electric current to flow through. The easiest is actually wire (T makes a circular motion with her hand). The easiest for the electric current to flow through is the wire. The moment you have bulbs, you have switches, you have other things eh, other components eh, it's actually more difficult. Some of you tried with the fan right? For the electric

current to flow through the fan is also more energy. And electric current, they will flow through the easiest way. So they always look for wires first. Are you clear?

As the extract reveals, T1 draws on domain-specific norms to emphasise the type and number of bulbs and switches, and not the colour, size or pattern (Turn 3). She highlights the need to be "very specific" in providing responses in the discipline of science. The teacher's justification triggers a clarifying question (Turn 6) and subsequently, her explanation revolves around the concept of resistance and the principle of the electric current flowing through "the easiest way" (Turn 7). Arguably, the student's question constitutes feedback to the teacher about the student's understanding about circuits, which T1 scaffolds via explanatory talk.

References

Abell SK, Volkmann MJ (2006) Seamless assessment in science: a guide for elementary and middle school teachers. Hainemann

Ajjawi R, Boud D (2017) Researching feedback dialogue: an interactional analysis approach. Assess Evaluat Higher Educ 42(2):252–265

Alexander RJ (2004) Towards dialogic teaching: rethinking classroom talk. Dialogos

Alonzo AC, Berry A, Nilsson P (2019) Unpacking the complexity of science teachers' PCK in action: enacted and personal PCK. In: Hume A, Cooper R, Borowski A (eds) Repositioning pedagogical content knowledge in teachers' knowledge for teaching science. Springer, New York

Anderson PJ, Peck CA (2007) Reconstructing practice in teacher education through artmaking. Read Matrix 7(1):19–29

Atwood S, Turnbull W, Carpendale JIM (2010) The construction of knowledge in classroom talk. J Learn Sci 19(3):358–402

Barnes D (2008) Exploratory talk for learning. In: Mercer N, Hodgkinson S (eds) Exploring talk in school: inspired by the work of Douglas Barnes. Sage, New York

Bell B, Cowie B (2001) Formative assessment and science education. Kluwer

Berk D, Hiebert J (2009) Improving the mathematics preparation of elementary teachers, one lesson at a time. Teach Teach Theory Pract 15(3):337–356

Black P, Harrison C (2001) Feedback in questioning and marking: the science teacher's role in formative assessment. School Sci Rev 82(301):55–61

Black PJ, Wiliam D (1998) Assessment and classroom learning. Assess Educ Principles Policy Pract 5(1):7–74

Blumenfeld P, Mergendoller J, Swarthout D (1987) Task as a heuristic for understanding student learning and motivation. J Curricul Stud 19(2):135–148

Boney K (2014) Beyond the skilled application of know-how: pedagogical reasoning as phronesis in highly competent teachers [Doctoral dissertation, University of South Florida]. Scholar Commons. http://scholarcommons.usf.edu/etd

Borg S (2006) The distinctive characteristics of foreign language teachers. Lang Teach Res 10(1):3–31

Brown GTL (2019) Is assessment for learning really assessment? Front Educ 4:64

Bruner JS (1986) Actual minds, possible worlds. Harvard University Press, Cambridge

Buck GA, Trauth-Nare AE (2009) Preparing teachers to make the formative assessment process integral to science teaching and learning. J Sci Teach Educ 20(5):475–494

Butler D, Winne P (1995) Feedback and self-regulated learning: a theoretical synthesis. Rev Educ Res 65(3):245–274

Bybee RW (1997) Achieving scientific literacy: from purposes to practices. Heinemann

Carless D, Boud D (2018) The development of student feedback literacy: enabling uptake of feedback. Assess Evaluat Higher Educ 43(8):1315–1325

Carlson J, Daehler KR, Alonzo AC, Barensden E, Berry A, Borowski A, Wilson CD (2019) The refined consensus model of pedagogical content knowledge in science education. In: Hume A, Cooper R, Borowski A (eds) Repositioning pedagogical content knowledge in teachers' knowledge for teaching science. Springer, New York, pp 77–92

Cazden C (1988) Classroom discourse: the language of teaching and learning. Heinemann

Chan PE, Konrad M, Gonzalez V, Peters MT, Ressa VA (2014) The critical role of feedback in formative instructional practices. Intervent School Clinic 50(2):96–104

Chin C (2006) Classroom interaction in science: teacher questioning and feedback to students' responses. Int J Sci Educ 28(11):1315–1346

Clark C, Yinger R (1979) Teachers thinking. In: Peterson P, Walberg H (eds) Research on teaching. McCutchan Press

Clarke S (2008) Active learning through formative assessment. Hodder Education

Coffey JE, Hammer D, Levin DM, Grant T (2011) The missing disciplinary substance of formative assessment. J Res Sci Teach 48(10):1109–1136

Datnow A, Hubbard L (2015) Teachers' use of assessment data to inform instruction: lessons from the past and prospects for the future. Teach College Rec 117:1–45

Datnow A, Lockton M, Weddle H (2021) Capacity building to bridge data use and instructional improvement through evidence on student thinking. Stud Educ Evaluat 69:100869

DeLuca C, LaPointe-McEwan D, Luhanga U (2016) Approaches to classroom assessment inventory: a new instrument to support teacher assessment literacy. Educ Assess 21(4):248–266

DeLuca C, Willis J, Cowie B, Harrison C, Coombs A, Gibson A et al (2019) Policies, programs and practices: exploring the complex dynamics of assessment education in teacher education across four countries. Front Educ 4:132

Duschl RA (2003) Assessment of inquiry. In: Atkin JM, Coffey JE (eds) Everyday assessment in the science classroom. National Science Teachers Association Press, pp 41–59

Earle S (2021) Formative decision-making in response to primary science classroom assessment: what to do next? Front Educ 5:584200

Embury-Dennis T (2020) Elon Musk predicts human language will be obsolete in as little as five years: 'we could still do it for sentimental reasons'. Independent. https://www.independent.co.uk/tech/elon-musk-joe-rogan-podcast-language-neuralink-grimes-baby-a9506451.html

Eraut M (2004) Informal learning in the workplace. Stud Cont Educ 26(2):247–273

Fisher D, Frey N, Hattie J, Thayre M (2017) Teaching literacy in the visible learning classroom: 6–12 classroom companion to visible learning for literacy. Sage, New York

Flavell JH (1979) Metacognition and cognitive monitoring: a new area of cognitive-developmental inquiry. Am Psychol 34(10):906–911

Freebody P (2004) Qualitative research in education: interaction and practice. Sage, New York

Galton MJ (2007) Learning and teaching in the primary classroom. Sage, New York

Garrison DR, Akyol Z (2015) Toward the development of a metacognition construct for communities of inquiry. Internet Higher Educ 24:66–71

Gillies RM (2012) Promoting reasoning, problem-solving and argumentation during small group discussions. In: Gillies RM (ed) Pedagogy: new developments in the learning sciences. Nova Science Publishers, pp 131–150

Good T, Brophy J (1987) Looking in classrooms. Harper & Row

Ha XV, Murray JC (2021) The impact of a professional development program on EFL teachers' beliefs about corrective feedback. System 96:102405

Hardman F (2008) Opening-up classroom discourse: the importance of teacher feedback. In: Mercer N, Hodgkinson S (eds) Exploring talk in school. Sage, New York, pp 131–150

Hattie J (2009) Visible learning: a synthesis of over 800 meta-analyses relating to achievement. Routledge, London
Hattie J (2012) Know thy impact. Feedback Learn 70(1):18–23
Hattie J (2019) The power of in-lesson verbal feedback. In: Hattie J, Clarke S (eds) Visible learning feedback. Routledge, London, pp 81–122
Hattie J, Timperley H (2007) The power of feedback. Rev Educ Res 77(1):81–112
Haug BS, Odegaard M (2015) Formative assessment and teachers' sensitivity to student responses. Int J Sci Educ 37(4):629–654
Hogan D, Chan M, Rahim R, Kwek D, Aye KM, Loo SC, Sheng YZ, Luo W (2013) Assessment and the logic of instructional practice in secondary 3 English and mathematics classrooms in Singapore. Rev Educ 1(1):57–106
Hussain FA (2022) Student feedback practices in the English Language classroom in Singapore: teacher pedagogical reasoning in task (re)design to 'ladle differently' [Doctoral dissertation, National Institute of Education, Nanyang Technological University]. NIE Digital Repository. http://hdl.handle.net/10497/24178
Khalaf BK (2018) Traditional and inquiry-based learning pedagogy: a systematic critical review. Int J Instr 11(4):545–564
Kidman G, Casinader N (2017) Inquiry-based teaching and learning across disciplines: comparative theory and practice in schools. Springer, New York
Kim M, Tan AL, Talaue FT (2013) New vision and challenges in inquiry-based curriculum change in Singapore. Int J Sci Educ 35(2):289–311
Korthagen FAJ (2010) Situated learning theory and the pedagogy of teacher education: towards an integrative view of teacher behavior and teacher learning. Teach Teach Educ 26(1):98–106
Kwek D, Lee Y-J, Wong HM, Miller R (2018) Core research programme: baseline investigation of science pedagogy (final report). National Institute of Education, Centre for Research in Pedagogy and Practice, Singapore
Kwek D (2023) [Lesson videos] [Unpublished raw data]. National Institute of Education, Centre for Research in Pedagogy and Practice, Singapore
Lee Y-A (2007) Third position in teacher talk: contingency and the work of teaching. J Pragmat 39(1):180–206
Lee I, Coniam D (2013) Introducing assessment for learning for EFL writing in an assessment of learning examination-driven system in Hong Kong. J Second Lang Writing 22(1):34–50
Lemke JL (1990) Talking science: language, learning and values. Ablex
Looney A, Cumming J, van Der Kleij F, Harris K (2018) Reconceptualising the role of teachers as assessors: teacher assessment identity. Assess Educ Principles Policy Pract 25(5):442–467
Loughran J, Berry A, Mullhall P (2012) Understanding and developing science teachers' pedagogical content knowledge, 2nd edn. Sense Publishers, Rotterdam
Mandinach EB, Schildkamp K (2021) Misconceptions about data-based decision making in education: an exploration of the literature. Stud Educ Evaluat 69:100842
Mansilla VB (2005) Assessing student work at disciplinary crossroads. Change Mag Higher Learn 37(1):14–21
Mercer N, Howe C (2012) Explaining the dialogic processes of teaching and learning: the value and potential of sociocultural theory. Learn Cult Soc Interact 1(1):12–21
Mercer N (1995) The guided construction of knowledge: talk amongst teachers and learners. Multilingual Matters
Ministry of Education (MOE) (2014) Science syllabus: primary. Ministry of Education, Singapore
Ministry of Education (MOE) (2020) Skills future for educators. Ministry of Education, Singapore
Minner DD, Levy AJ, Century J (2010) Inquiry-based science instruction: what is it and does it matter? Results from research synthesis years 1984 to 2002. J Res Sci Teach 47(4):474–496
Mortimer E, Scott P (2003) Analysing discourse in the science classroom. In: Millar R, Leach J, Osborne J (eds) Improving science education: the contribution of research. Open University Press

Mourshed M, Chijioke C, Barber M (2010) Education: how the world's most improved school systems keep getting better. McKinsey
Nassaji H, Wells G (2000) What's the use of 'Triadic dialogue'? An investigation of teacher-student interaction. Appl Linguist 21(3):376–406
Nystrand M, Gamoran A (1991) Instructional discourse, student engagement, and literature achievement. Res Teach English 25(3):261–290
Nystrand M, Wu A, Gamoran A, Zeiser S, Long DA (2003) Questions in time: investigating the structure and dynamics of unfolding classroom discourse. Discourse Process 35(3):135–198
Popham WJ (2008) Transformative assessment. Association for Supervision and Curriculum Development
Price M, Handley K, Millar J (2011) Feedback: focusing attention on engagement. Stud Higher Educ 36(8):879–896
Ratnam-Lim C, Tan KHK (2015) Large-scale implementation of formative assessment practices in an examination-oriented culture. Assess Educ Principles Policy Pract 22(1):61–78
Rubie-Davies CM, Flint A, McDonald LG (2012) Teacher beliefs, teacher characteristics, and school contextual factors: what are the relationships? Br J Educ Psychol 82(2):270–288
Ruiz-Primo MA, Furtak EM (2007) Exploring teachers' informal assessment practices and students' understanding in the context of scientific inquiry. J Res Sci Teach 44(1):57–84
Sadler DR (1998) Formative assessment: revisiting the territory. Assess Educ 5(1):77–84
Seah KKC (2016) PSLE changes: mixed effects on pupils of different abilities. The Straits Times. https://www.straitstimes.com/opinion/mixed-effects-on-pupils-of-different-abilities
Seedhouse P (2005) "Task" as research construct. Lang Learn 55(3):533–570
Seung E, Bryan LA, Haugan MP (2012) Examining physics graduate teaching assistants' pedagogical content knowledge for teaching a new physics curriculum. J Sci Teach Educ 23(5):451–479
Shavelson RJ, Cadwell J, Izu T (1977) Teachers' sensitivity to the reliability of information in making pedagogical decisions. Am Educ Res J 14(2):83–97
Shepard L (1995) Using assessment to improve learning. Educ Lead 52(5):38–43
Shulman L (1987) Knowledge and teaching: foundations of the new reform. Harvard Educ Rev 57(1):1–22
Shute V (2008) Focus on formative feedback. Rev Educ Res 78(1):153–189
Sinclair J, Coulthard M (1975) Towards an analysis of discourse. Oxford University Press, Oxford
Stein MK, Grover B, Henningsen M (1996) Building student capacity for mathematical thinking and reasoning: an analysis of mathematical tasks used in reform classrooms. Am Educ Res J 33(2):455–488
Stiggins R, DuFour R (2009) Maximizing the power of formative assessments. Phi Delta Kappan 90(9):640–644
Storch N, Wigglesworth G (2010) Learners' processing, uptake and retention of corrective feedback on writing: case studies. Stud Second Lang Acquis 32(2):303–334
Tan P-H, Tan A-L (2011) Assessing students' learning of primary science in the multicultural context of Singapore: considerations influencing task selection for formative assessment [Paper presentation]. In: Proceedings of the 37th annual conference of the international association of educational assessment (IAEA), Manila
Tan YK, Chow HH, Goh C (2008) Examinations in Singapore: change and continuity (1891–2007). World Scientific
Tan JP, Koh E, Chan M, Costes-Onishi P, Hung D (2017) Advancing 21st century competencies in Singapore. Asia Society: Center for Global Education. https://asiasociety.org/sites/default/files/2017-10/advancing-21st-century-competencies-in-singapore.pdf
Tan P-H, Tan A-L (2014) Teacher's ideas and concerns with assessment practices in inquiry science. In: Tan A-L, Poon C-L, Lim SSL (eds) Inquiry into the Singapore science classroom: research and practices. Springer, New York, pp 67–87

Tan KHK, Wong HM (2018) Assessment feedback in primary schools in Singapore and beyond. In: Lipnevich AA, Smith JK (eds) The Cambridge handbook of instructional feedback. Cambridge University Press, Cambridge, pp 123–144

Tay HY, Lam KWL (2022) Students' engagement across a typology of teacher feedback practices. Educ Res Policy Pract 21:427–445

To J (2022) Using learner-centred feedback design to promote students' engagement with feedback. Higher Educ Res Develop 41(4):1309–1324

To J, Tan K, Lim M (2023) From error-focused to learner-centred feedback practices: unpacking the development of teacher feedback literacy. Teach Teach Educ 131:104185

Torrance H, Pryor J (1998) Investigating formative assessment: teaching, learning and assessment in the classroom. Open University Press

van Zee EH, Iwasyk M, Kurose A, Simpson D, Wild J (2001) Student and teacher questioning during conversations about science. J Res Sci Teach 38(2):159–190

Vygotsky LS (1978) Mind in society: the development of higher psychological processes. Harvard University Press, Harvard

Wiliam D (2006) Assessment: learning communities can use it to engineers a bridge connecting teaching and learning. J Staff Develop 27(1):16–20

Zhai J, Tan A-L (2015) Roles of teachers in orchestrating learning in elementary science classrooms. Res Sci Educ 45(6):907–926

Chapter 9
Integrated Science, Technology, Engineering, and Mathematics—Problem-Based Learning—Education for Sustainable Development (I-STEM-PBL-ESD) Framework

Aaron A. Funa, Lydia S. Roleda, and Maricar S. Prudente

9.1 Introduction

Education for sustainable development (ESD) is a concept that was designed to analyze and improve educational institutions in a systematic manner, encouraging and energizing sustainability in the minds, hearts, and actions of future generations (Zguir et al. 2021). It equips learners with the knowledge, skills, values, and attitudes necessary to make sound judgments and conduct responsible actions in the interests of environmental integrity, economic viability, and fair society (United Nations Educational, Scientific, and Cultural Organization [UNESCO] 2018). It is recognized as the key enabler for the 2030 Sustainable Development Goals (SDGs) and a critical component of quality education and student empowerment (Laurie et al. 2016; UNESCO 2018). However, as the world advances, so do educational expectations. As a result, several research studies have proposed improvements to ESD implementation to cope with these advancements (Ekantini and Wilujeng 2018; Michalos et al. 2017; Olsson et al. 2015). These advancements respond to UNESCO's urgent request to scale up sustainable development (SD) interventions and fulfill changing educational expectations (UNESCO 2020). In this regard, the current researchers contribute to this effort by designing an instructional framework that would help

A. A. Funa (✉)
Sorsogon State University, Sorsogon, Philippines
e-mail: funa.aaron@sorsu.edu.ph

A. A. Funa · L. S. Roleda · M. S. Prudente
Department of Science Education, Br. Andrew Gonzalez FSC College of Education, De La Salle University, Manila, Philippines

Y. S. Ong et al. (eds.), *A Diversity of Pathways through Science Education*,
https://doi.org/10.1007/978-981-97-2607-3_9

teachers in planning their instructional approach toward integrating and promoting SD in education.

Utilizing Torraco's (2005) guide for conducting integrative literature reviews, the researchers examined the existing literature addressing the following research inquiries:

1. What is ESD?
2. How is the Problem-Based Learning (PBL) approach employed within ESD?
3. What are the key components and principles of the I-STEM (Integrated Science, Technology, Engineering, and Mathematics) approach?
4. How is the I-STEM-PBL instructional approach designed and implemented?
5. In what ways is STEM integrated into the context of ESD?
6. What insights and applications does the I-STEM-PBL-ESD instructional framework offer in the field of ESD?

9.1.1 Definition of ESD

The United Nations Division for Sustainable Development (UNDSD 1992) reaffirmed the notion that ESD is a critical facilitator for SD. They stressed at the United Nations Conference on Development and the Environment in Rio de Janeiro, Brazil, that ESD, particularly for children, is essential for the long-term success of Agenda 21—a comprehensive plan of action that has been implemented globally, nationally, and locally to address concerns relating to human-induced environmental degradation (UNDSD 1992). According to the Conference, education's role in the SD is to help the youth develop their distinctive viewpoints, participate in making crucial decisions, and defend themselves from harm brought about by environmental damage.

The growing acknowledgment of ESD as a necessary component of quality education and a vital element for SD necessitates improvements in its implementation policies and initiatives (Laurie et al. 2016; UNESCO 2014, 2018). In particular, SDG4 (Quality Education) aspires to guarantee that all learners possess the required knowledge and skills for SD by 2030. These competencies and knowledge refer to sustainable lifestyles, human rights, gender equality, a culture of peace and nonviolence, global citizenship, and a respect for cultural variety and culture's contribution to SD (UNESCO 2018). These goals made the SDG4 necessary for the effective attainment of the other SDGs, just as ESD is important for obtaining quality education (SDG4).

ESD is a participatory approach that empowers students to enhance their knowledge, attitudes, and behaviors (KAB) related to SD, including climate change, disaster risk reduction, biodiversity, and poverty reduction (Leicht et al. 2018). It fosters collaborative learning, critical and future-oriented thinking, and sustainable consumption (UNESCO 2011), enabling individuals of all ages to take informed and responsible actions for SD (UNESCO 2014). ESD is interdisciplinary and transdisciplinary, accommodating contributions from various disciplines (Leicht et al. 2018). It emphasize the improvement of students' conceptual understanding, creativity, and KAB to promote SD and address its key dimensions: economic development, social inclusion, Earth's physical systems, and good governance (Klarin 2018; Sachs 2015). Unlike other pedagogies, ESD places a strong focus on achieving Sustainable Development Goals (SDGs).

9.1.1.1 The Learner-Centered Action-Oriented Transformative Pedagogy

UNESCO (2014) identified five key areas for prioritizing ESD: policy advancement, the transformation of learning environments, educator and trainer capacity building, youth empowerment and mobilization, and the acceleration of sustainable solutions. To effectively address these priorities, a learner-centered, action-oriented, and transformative pedagogical approach, as outlined by UNESCO (2017), was recommended. The learner-centered aspect emphasizes students' autonomy in learning, active engagement in knowledge construction, free expression of ideas, and personal reflection on their decisions and actions (Doyle 2011; Rieckmann 2018; Weimer 2013). Action-oriented learning encourages students to apply their knowledge and reflect on real-life experiences gained from community projects, internships, immersions, and field exposure activities (Rieckmann 2018; Suškevičs et al. 2019). This involves concrete experiences, observations, reflections, the formation of abstract concepts, and the application of knowledge, skills, and practices in novel situations (Kolb 1984; Morris 2020). Transformative learning empowers students to critically evaluate and change their perceptions and behaviors in response to various situations (Slavich and Zimbardo 2012). This approach equips them to challenge the status quo, embrace disruptive thinking, and construct new knowledge based on prior concepts and present experiences (Lotz-Sisitka et al. 2015; Rieckmann 2018).

According to Rieckmann (2018), the learner-centered action-oriented transformative pedagogy can be manifested by the collaborative real-world projects (e.g., immersion projects and campaigns), vision-building exercises (e.g., future creating workshops), analysis of complex systems (e.g., community-based research and case studies), and critical and reflective thinking (e.g., fishbowl discussion and reflective journals). Consequently, the current researchers determined approaches to facilitate carrying out this pedagogy and integrate real-world problems in education in the form of PBL and I-STEM approaches.

9.1.2 PBL as Employed for ESD

The implementation of ESD may be facilitated by the use of PBL. PBL is characterized as a social constructivist approach to education that emphasizes active learning, problem-solving, and 21st-century skills, as well as self-regulation, inquiry, and metacognition applicable to solving real-world problems (Barrows 1996; Grant and Tamim 2019). Over the last five decades, numerous authors have defined PBL in a variety of ways, depending on their intended audience. PBL may be defined as a process of inquiry, learning to learn, and building mental models of the world, the latter of which constitutes a "cognitive constructivist" approach that has resulted in multiple research studies (Schmidt et al. 2019, 2009). Six defining characteristics of PBL have been identified based on these three perspectives: (1) learning through real-world problems; (2) small group formation collaboration; (3) teacher/tutor guidance; (4) limited number of lectures; (5) student-initiated learning; and (6) ample time available to study and work with the problems (Evensen and Hmelo 2000; Hmelo-Silver 2004; Schmidt 1983, 1993). PBL has a strong and beneficial influence on students' academic achievement, with students scoring better on all examinations than the control group that uses traditional/conventional methods (Dochy et al. 2003; Funa and Prudente 2021).

PBL originated in 1969 at McMaster University in Ontario, Canada, where it was introduced as an innovative curriculum. Initially, twenty medical students participated in a self-directed learning approach guided by teacher-created problems (Hamilton 1976; Neufeld and Barrows 1974). These students worked in small groups and progressed through four phases to graduate. Phase I introduced students to the medical community and its foundations, emphasizing group collaboration. Phase II focused on fundamental concepts. In Phase III, a variety of PBL methods encouraged students to explore specific topics, while lectures and demonstrations were limited to four or five per week. Phase IV involved students in hands-on patient care, requiring them to address patients' needs comprehensively. As students advanced through the phases, their roles became increasingly complex.

In the PBL curriculum, the core element is the collection of problems (Barrows 1996). These problems are deliberately designed to be open-ended and lack a single formulaic answer, encouraging students to inquire and seek additional information (Savery 2019). Despite their inherent complexity, it's essential to craft these problems intelligently, as they serve several crucial purposes: fostering problem-solving skills, promoting in-depth study of topic information, connecting abstract knowledge to practical application, immediate use of content knowledge, and motivating active learning (Hung 2019). In contrast to lecture-based education, the challenges faced by teachers when implementing PBL revolve around formulating high-quality, effective, and comprehensive problems. These problems should not merely serve as an initial hook to engage students; they should be an integral part of the learning process, driving knowledge, attitude, and behavior development throughout the course. Therefore, it is imperative to ensure that the problems used in the PBL curriculum are well-constructed.

The nature of problems used in PBL can significantly influence outcomes (Duch et al. 2001; Hung et al. 2019; Sockalingam and Schmidt 2011). PBL offers several advantages, including promoting in-depth understanding, fostering reasoned decision-making, connecting prior knowledge to new concepts, and encouraging collaboration. However, poorly designed PBL problems can lead to concerns. First, PBL may reduce the time required to cover course materials compared to traditional teaching methods (Albanese and Mitchell 1993). On average, students in PBL grasp about 50% of the course content, whereas lecture-based methods achieve a 100% coverage rate (Hung 2019). Second, selecting and contextualizing real-world problems can be challenging and resource-intensive, especially in non-professional courses (Hung 2019). Third, motivating students through problems can be difficult, with initial enthusiasm possibly diminishing over time (Moust et al. 2005; Romito and Eckert 2011), despite the initial improvement in their intrinsic and extrinsic motivation for subject-specific knowledge acquisition.

To address the challenges associated with PBL problems, including the creation of high-quality problems, time management, and student engagement, several guidelines have been proposed. Duch et al. (2001) and Torp and Sage (2002) have outlined essential steps for planning and creating effective PBL problems. These guidelines include selecting a central topic aligned with standards and curriculum, specifying clear learning objectives to guide problem formulation, considering real-world contexts and student interests, introducing problems gradually in stages, and providing a teacher's guide for smoother implementation. These strategies aim to mitigate issues related to problem quality, time constraints, and student motivation in PBL. Additionally, Hung (2006) introduced the 3C3R PBL problem design model, which encompasses core and processing components. The core components focus on content, context, and connection, ensuring that students acquire discipline-based knowledge, the context aligns with the curriculum, and problems align with lesson goals. The processing components, including researching, reasoning, and reflecting, aim to enhance cognitive processing and decision-making skills through inquiry-based learning, learning to learn, and problem-solving. Notably, in 2019, the 3C3R model was further refined, resulting in the development of the second-generation 3C3R PBL problem design model, which continues to advance PBL in education.

9.1.2.1 The Second Generation 3C3R PBL Model

The second generation of the 3C3R PBL problem design model, introduced by Hung (2019), is designed to aid instructional designers and educators in creating high-quality, effective problems for PBL teaching across various student levels and disciplines. This model addresses students' behaviors, perceived difficulties, and teamwork in the context of PBL instruction. It expands on the core and processing components by incorporating affect, difficulty, and teamwork elements, ensuring that problems engage, motivate, and address students' psychological, emotional, and social needs. The second generation 3C3R model is informed by empirical investigations and PBL insights from previous work by Hung (2006, 2009, 2019), making it a valuable guide for developing PBL problems. This model not only assesses students' discipline-specific knowledge and problem-solving skills but also considers their attitudes and emotional disposition toward learning, nurturing the three essential pillars of attitudes, skills, and knowledge critical for students' future professional development (Bakarman 2005).

While selecting the right model for crafting quality problems is crucial in PBL education, models for implementing PBL processes are equally important, especially in basic and tertiary education settings. Notable models include Barrows and Myers'

Secondary Education Model (Barrows and Myers 1993) and the Center for Problem-Based Learning's K-16 PBL Model (Torp and Sage 2002), in which the latter has broader applicability to various learner levels.

9.1.2.2 The K-16 PBL Model by the Center for PBL

Torp and Sage (2002) developed the K-16 PBL Model, which consists of design and implementation processes tailored for PBL instruction. The K-16 educational system includes basic education (grades 1–12) and four years of post-secondary education (grades 13–16). These processes include preparing learners, introducing the problem, identifying knowledge gaps, defining the problem, gathering and sharing information, generating solutions, determining the best-fit solution, presenting the solution, and debriefing the problem. The third, fourth, and fifth steps are iterative, with the teacher's continuous involvement in coaching and evaluating students.

In the Philippines, the government, through Republic Act No. 10533, mandates the use of constructivist methodologies like PBL in education (DepEd 2019a). The Alternative Delivery Mode Learning Resource Standards (ADMLR Standards) version 12 of DepEd in 2019 also promotes constructivist and inquiry-based pedagogical approaches, including PBL, in response to the COVID-19 pandemic (DepEd 2019b). Furthermore, PBL is employed to enhance scientific, technical, and environmental literacy at the basic education level (DepEd 2016). However, to maximize its effectiveness and meet educational objectives, PBL needs to be complemented with other teaching methodologies and scaffolding approaches (Funa and Prudente 2021). In this study, an integrated STEM approach was used to complement PBL methods, incorporating the 2nd generation 3C3R problem-based model for designing high-quality PBL problems and the K-16 PBL Model for scaffolding PBL instruction.

9.1.3 Key Components and Principles of Integrated Stem Approach

STEM has garnered widespread attention throughout the years due to its potential to elevate the prestige and quality of science and mathematics education around the globe (Honey et al. 2014). STEM was initially coined by the National Science Foundation in the United States as a way to identify and connect the four domains (S-T-E-M) and was subsequently used extensively in policies and curricula to refer to educational outcomes relevant to economies, technological innovations, 21st-century skills, and national security (Bybee 2013). According to McComas and Burgin (2020), there are two prominent definitions of how STEM is being used in education. On one hand, it is defined as a field of study that focuses on any of its four components (S, T, E, or M), which means that a math teacher and a science teacher are most certainly STEM educators. On the other hand, it is described as a group of domains that necessitates the presence of several or all of its components (S + T + E + M), as required to obtain the STEM imprimatur.

STEM education enhances students' understanding of how things work and their ability to use technology. It introduces engineering, fosters problem-solving, and encourages innovation. STEM involves group activities, laboratory work, and projects, developing critical 21st-century skills and promoting decision-making in areas like health, energy efficiency, and environmental sustainability (Bybee 2010). It's seen as a way to develop innovative thinkers for the future workforce, supporting the economy and sustainable development (Marrero et al. 2014; Tan et al. 2019).

Despite its potential, STEM education lacks a clear, shared definition among educators and faces challenges in integration (Honey et al. 2014; Kelley and Knowles 2016). The breadth of STEM complexity poses difficulties in comprehension, framework development, implementation, and evaluation (Koh 2018). To simplify, McComas and Burgin (2020) suggest three ways to blend STEM elements: transdisciplinary, integrative, and integrated. Transdisciplinary applies knowledge from multiple STEM areas to real-world problems, integrative uses engineering methods to teach science and math simultaneously with technology/engineering education, and integrated combines any or all STEM elements (McComas and Burgin 2020). To clarify these terms, the present researchers use "I-STEM" to distinguish teaching with two or more STEM elements from teaching with any single element.

Over time, STEM education has made significant contributions, but educators and students still need to fully harness its opportunities (Koh 2018). STEM must focus on developing students' abilities to understand and address complex real-world challenges, adopting an I-STEM approach in their daily lives. In this study, "integrated" is used alongside STEM to emphasize the interconnectedness of the four domains, adopting the term "I-STEM" coined by McComas and Burgin (2020). El Sayary et al. (2015) discuss two approaches to I-STEM education: teaching each discipline with a focus on two or more subjects (SteM), and combining three domains into one (E $\rightarrow$ STM).

To comprehend I-STEM and its various interpretations, the researchers analyzed different definitions. Sanders (2009) described I-STEM as an approach involving two or more STEM areas or connections between STEM and other subjects. Kelley and Knowles (2016) defined it as teaching STEM content from multiple domains, fostering connections and practices in an authentic context. Bryan et al. (2016) viewed it as learning science and math through integrating engineering and design practices. Tan et al. (2019) saw it as blurring the boundaries of the four domains for mutual enhancement to meet the demands of the fourth industry. Considering these definitions, the current researchers viewed each domain (science, technology, engineering, and mathematics) as distinct but interconnected fields. Science deals with empirical exploration and explanation, technology involves resourceful use of tools, engineering focuses on systematic problem-solving, and mathematics provides essential measurements. These domains, when combined, yield contextualized solutions to problems, not limited to STEM fields. However, this study mainly explains how I-STEM is incorporated with ESD. Expanding I-STEM into non-STEM fields is recommended.

By integrating I-STEM into education, students gain a holistic perspective for problem-solving, applying discipline-based knowledge to real-world sustainability issues, both within and beyond STEM. These simplified definitions of I-STEM aim to facilitate its use, primarily in K-16 education, without replacing existing STEM or I-STEM definitions. They were derived from the core concepts of each STEM domain and serve to teach topics and enable students to address real-world sustainability challenges.

9.1.4 I-STEM-PBL Instructional Approach

Teachers often use ill-structured problems in PBL to prepare students for real-world challenges, as these situations lack clear structures (El Sayary et al. 2015). The complexity of PBL problems correlates with increased skill acquisition (Capraro et al. 2013; Russell et al. 2007; Lopatto 2004). Barrows (1986) classified PBL into six levels based on issue structure and self-directedness, ranging from lecture-based with problem-solving to pure PBL. Pure PBL, identified by Harmer and Cates (2007) as the most effective for real-world, transdisciplinary STEM situations.

PBL significantly boosts students' interest in STEM-related careers (LaForce et al. 2017). Combining PBL with I-STEM enhances higher-order thinking skills, including cognitive (critical thinking and creativity), content (interdisciplinary, technology, and connections), and collaborative learning (collaboration, communication, and self-direction) (El Sayary et al. 2015). The present researchers believe that the synergy between PBL and I-STEM can improve conceptual comprehension, creativity, and problem-solving abilities, creating a more robust scientific teaching and learning approach.

9.1.5 STEM for ESD

The integration of STEM into ESD has been a prominent subject of rigorous research and scholarly exploration. Researchers such as Suh and Han (2019) and Fakhrudin et al. (2021) have delved into this integration, showcasing its application in addressing real-world sustainability challenges. However, the discussion on the intricate relationships between diverse STEM domains in the context of solving real-life sustainability problems in these studies are limited. Additionally, instructional frameworks employed needs to be highlighted.

Conversely, some studies have focused on integrating ESD into STEM learning. For example, Hopkinson and James (2010) conducted a comprehensive review, shedding light on recent instances of embedding ESD within science and related curricula. Their work underscores the importance of creating meaningful and relevant educational experiences for both educators and students, encompassing various embedding strategies and discourses. Nonetheless, in contrast to the aforementioned studies, the present study introduces an instructional framework that demonstrates how discipline-based knowledge within STEM can be applied to address real-world challenges, particularly in for ESD. The I-STEM-PBL-ESD instructional framework offers another perspective on teaching students how to utilize STEM concepts and methodologies to tackle daily problems and contribute to sustainable solutions. Consequently, it serves as a tool for educators seeking innovative approaches to empower students with practical STEM skills in their daily lives.

9.2 I-STEM-PBL-ESD: The Instructional Framework

Historically, PBL has a strong and more beneficial influence on students' academic achievement than traditional/conventional methods (Dochy et al. 2003; Funa and Prudente 2021). Hence, the current researchers developed a framework that will guide teachers in creating and contextualizing problems for PBL as well as in implementing a PBL approach in the learning and teaching processes. They included the concepts of the 2nd generation 3C3R problem-based design and the K-16 PBL model, as well as incorporating the four STEM domains into solving real-world sustainability concerns. The inclusion of I-STEM in the framework is important to contribute to a more unified understanding of the definition and impact of STEM education, calling the framework the I-STEM-PBL-ESD Instructional Framework. The I-STEM-PBL-ESD Instructional Framework consists of three parts: the development and implementation of I-STEM-PBL modules, the student learning, and the output (see Fig. 9.1).

9.2.1 STEM-PBL Module

Parameters such as learner-centered action-oriented transformative pedagogy, PBL elements, and I-STEM approach to learning and teaching were considered in the development of the I-STEM-PBL-ESD instructional framework. The module for I-STEM-PBL instruction is composed of I-STEM and PBL approaches, with the former being adapted from Dugger and Fellow (2011), McComas and Burgin (2020), and Tan et al. (2019) as a mechanism for solving problems in PBL instruction using science, technology, engineering, and mathematics as distinct but interconnected domains; and the latter being adapted from the 2nd generation 3C3R problem design model by Hung (2019) for developing PBL problems and the K-16 PBL model by Torp and Sage (2002) for scaffolding PBL instructions. These approaches, when combined through I-STEM-PBL modules, may be used to increase students' meaningful conceptual understanding (MCU), cognitive process associated with creativity (CPAC), and knowledge, attitudes, and behaviors (KAB) towards SD, which may lead to the promotion of ESD. In turn, the MCU, CPAC, and KAB provide the input necessary to enhance the I-STEM-PBL modules and their implementation.

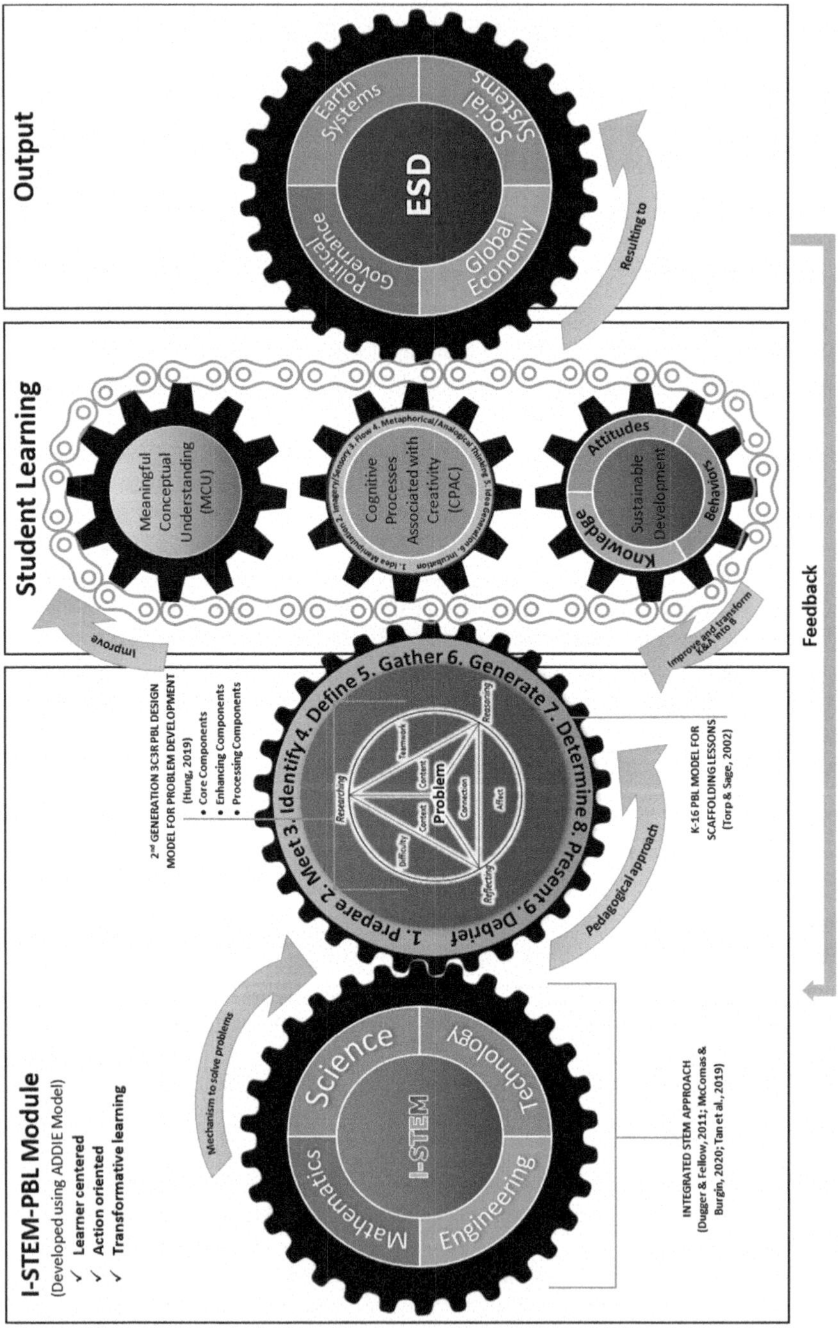

Fig. 9.1 The I-STEM-PBL-ESD instructional framework

The researchers adapted the models in the study of Dugger and Fellow (2011), McComas and Burgin (2020), and Tan et al. (2019) to merge the four distinct but interrelated fields of STEM. On the other hand, the 2nd generation 3C3R model by Hung (2019) and the K-16 PBL model by Torp and Sage (2002) were adapted to properly implement the PBL approach. The distinction, however, in this framework is that students are not expected to create a product and are not necessarily needed to work in the community unless they are collecting data. The main objective is to develop students' abilities for understanding and acting on complex real-world challenges, as well as for autonomously adopting an I-STEM approach as part of their daily lives while learning the content defined by the curriculum standards. It is crucial to clarify that this framework focuses on Problem-Based Learning (PBL), distinct from Project-Based Learning (PjBL). In PBL, students generate ideas and practice problem-solving skills in the classroom, with community visits mainly for data collection, not immediate implementation. This approach accommodates time constraints associated with curriculum learning. However, the study does stress the importance of later testing students' proposals, distinguishing it from PjBL, which requires immediate implementation. Nevertheless, students are encouraged to continue implementing and testing their ideas and outputs profoundly using the I-STEM-PBL-ESD framework in the future.

9.2.2 MCU, CPAC, and KAB: Student Learning

According to the World Education Forum (2015: para. 9), quality education nurtures creativity and knowledge while assuring the learning of core skills such as literacy and numeracy, as well as analytical, problem-solving, and other advanced cognitive, interpersonal, and social abilities. Additionally, through ESD and global citizenship education, quality education teaches the skills, values, and attitudes necessary for individuals to live healthy and fulfilling lives, make informed choices, and react to local and global concerns. In this sense, the present researchers recognized that among the important variables to assess and evaluate learning, the students' MCU, CPAC, and KAB towards SD play important roles in ESD.

9.2.2.1 Mutual Conceptual Understanding (MCU)

MCU is distinct from conceptual understanding alone. Roth (1990) described this concept as students' capacity to integrate past knowledge and domain knowledge to get a better grasp of the world. It discourages rote memorizing and instead encourages students to expand their knowledge by employing it in real-life situations (Roth 1990; Widiyatmoko and Shimizu 2018). Given this definition, it is critical to examine this variable among students in response to promoting SD.

Mayer's (1983) schema theory is one of the theoretical bases for MCU. This theory stipulates three requirements for learning to occur: (1) learners must be provided with materials; (2) they must possess relevant previous knowledge or schema; (3) they must activate that knowledge or schema. Extending it further will result in meaningful integration of learning, which Piaget (1929) and Posner et al. (1982) refer to as "assimilation" and Hewson and Hewson (1984) refer to as "integration," as well as meaningful conceptual change in learning, which Piaget (1929) and Posner et al. (1982) refer to as "accommodation" and Hewson and Hewson (1984) refer to as "reconciliation or exchange."

9.2.2.2 Cognitive Processes Associated with Creativity (CPAC)

The CPAC is concerned with the cognitive processes that occur during the students' creative processing activities (Miller 2014). It involves a variety of creative processes, including brainstorming, metaphorical and analogical thinking, perspective-taking, visual or sensory processing, incubation, and flow. Miller (2014) characterizes these processes as follows: These processes are characterized as follows: (a) Brainstorming involves generating numerous ideas, regardless of quality; (b) Metaphorical and analogical thinking transfers concepts across contexts to produce novel ideas; (c) Viewpoint-taking involves altering one's perspective for innovative solutions; (d) Sensory perception, including visualization and other sensory channels, plays a crucial role; (e) Incubation is a preconscious state allowing unconscious problem-solving; and (f) Flow refers to automatic, aware creativity during challenging tasks.

Along with the MCU, the researchers recognized the significance of investigating students' CPAC because creativity is a necessary component of quality education and 21st-century skills development. Knowing the students' creative processes may enable the researchers to understand and explain how students apply their discipline-based scientific knowledge to produce something original and worthwhile as they address real-world sustainability problems more fully.

9.2.2.3 Knowledge, Attitudes, and Behaviors (KAB) Towards SD

Teachers and students both contribute significantly to the effectiveness of ESD implementation. Both the teachers and their educational institutions are seen as critical change agents in terms of influencing and deepening students' commitment to SD throughout their formative years of school (McKeown and Hopkins 2017), while students of all ages are required to prepare for and be capable of confronting and resolving contemporary and future real-world problems (UNESCO 2018, 2020). To support and realize these functions, UNESCO (2020) urgently calls on concerned stakeholders and authorities to scale up interventions, enumerating the tasks that must be accomplished. These tasks include an awareness of the information, skills, beliefs, and attitudes that would enable informed judgments and responsible individual and communal actions to local, national, and global concerns while taking the complex systems and the future generations into account (UNESCO 2020).

In addition, the teachers' and students' association of knowledge, attitudes, and behaviors are different from each other, and they are having difficulties transforming favorable knowledge and attitudes into positive behaviors towards SD (Funa et al. 2022a, b). Given these ideas, assessing teachers' and students' KAB is critical for changing and aligning the learning environment with SD principles that would allow manifestations of positive behaviors and practices toward promoting SD. As a result, the researchers included the KAB as part of the variables for student learning because of their importance as a component of ESD.

9.2.3 Balancing the Four Interacting Complex Systems of ESD: The Output

SD encompasses not only economic development but also environmental sustainability. Elkington (1994) introduced the triple bottom line concept, emphasizing the interconnectedness of environmental, societal, and economic pillars in SD. To achieve SD comprehensively, all these pillars must be sustainable (Klarin 2018). Sachs (2015) further refined SD, highlighting the need to balance four interdependent complexes: the global economy, social systems, Earth systems, and political governance. This holistic approach emerged due to the limitations of the Millennium Development Goals from 2000 to 2015. While poverty decreased significantly, inequalities and environmental degradation increased, leading to the adoption of the 17 SDGs for 2015–2030 (Sachs 2015).

Education plays a vital role in advancing SD, instilling environmental responsibility and problem-solving skills. Bernice Goldsmith emphasized the role of education in teaching individuals their responsibilities in securing a sustainable future (WCED 1987). The WCED's 1987 report, "Our Common Future," highlighted education's roles in shifting values toward SD, enhancing problem-solving abilities, and increasing environmental participation. Training and supporting teachers globally, with a focus on centralized global objectives like the 17 SDGs, were deemed essential (WCED 1987).

In this study, under the I-STEM-PBL-ESD instructional framework and utilizing the I-STEM-PBL approach, the researchers aim to contribute to UNESCO's (2020) call to expand SD initiatives and address evolving educational needs. These needs extend beyond achieving SDGs to include preparedness for emergencies such as pandemics, wars, political challenges, and environmental issues while maintaining a balance among the four interconnected SD complexes (Sachs 2015).

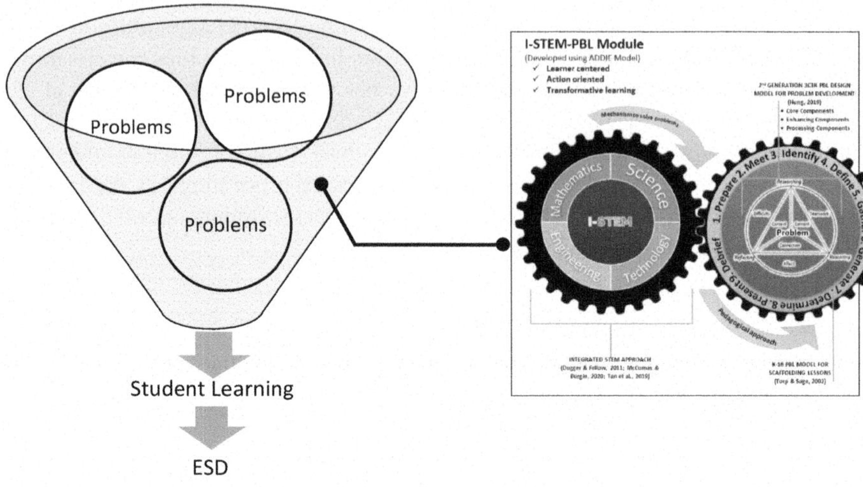

Fig. 9.2 Problems as drivers of learning

9.2.4 Hypothetical Implementation Scenario on the Use of I-STEM-PBL-ESD

Problems are at the core of PBL instruction. Hence, the present researchers leverage these problems to facilitate learning through the development and utilization of I-STEM-PBL learning modules, which are central to the I-STEM-PBL-ESD Instructional Framework (Fig. 9.2). The process begins by defining the system, evaluating the current situation, and analyzing the underlying causes of identified issues by synthesizing relevant literature and studies.

To illustrate how this framework can be applied, let's consider a hypothetical scenario involving the use of this framework by teachers. They could select problems related to Pili (*Canarium ovatum*), a tree endemic to the Bicol region in the Philippines, known for its nuts, oils, and elemi (Millena and Sagum 2018). Research has shown that Pili nuts can serve as alternatives to internationally recognized nuts like macadamia and almonds, its oil can be a cost-effective substitute for olive oil, and its elemi is used in luxury fragrances and pharmaceuticals (Catelo and Jimenez 2016). Despite its economic potential, the interest among youth in farming and cultivating Pili has declined, with many opting for white-collar jobs or careers as call-center agents or construction workers (Sorsogon Pili Board 2018; Sorsogon Pili Roadmap 2018). Recognizing the significance of integrating Pili's sustainable development into education, the researchers employ this framework.

Having identified the problem, the next step is to align it with the characteristics of the target students, educational trends, methods, and existing curriculum standards. For instance, the issue of Pili can be integrated into grades 10 and 11 Biology courses within the Philippine K-12 curriculum, which cover concepts related to organic molecules like carbohydrates, proteins, lipids, and nucleic acids, along with research courses that emphasize data collection and problem-solving.

With the essential variables in mind, teachers must formulate a high-quality PBL problem. These problems should pique students' interest in identifying biological molecules in Pili plants and using them to address real-life challenges. The teachers can follow the 2nd generation 3C3R problem design model (Hung 2009, 2019) to ensure core (content, context, and connection), processing (researching, reasoning, and reflecting), and enhancing (affect, difficulty, and teamwork) components are present. Moreover, the problem should catalyze learner-centered, action-oriented, and transformative learning.

After formulating the problem, it needs to be introduced to students in a scaffolded manner through tasks that gradually lead to its resolution. The I-STEM-PBL-ESD instructional framework, particularly the K-16 PBL model (Torp and Sage 2002), guides this process with nine steps: (1) preparing learners, (2) introducing the problem, (3) identifying prior knowledge and questions, (4) defining the problem statement, (5) collecting and sharing information, (6) generating possible solutions, (7) determining the best-fit solutions, (8) presenting the solution, and (9) debriefing the problem.

To solve the problem, students' solutions should encompass all four STEM domains, defining their roles, responsibilities, and limitations. Each small group within the class may have members with roles in specific STEM fields, such as a science expert, technology expert, engineering expert, mathematics expert, and a sustainable development expert. These roles ensure a holistic approach to solving the problem while considering the balance among the four interacting complex systems of economic development, social inclusion, Earth's physical systems, and good governance.

Once the I-STEM-PBL module is ready with all components in place, it can be implemented. To address time constraints, the implementation phase can be divided into parts without compromising the duration allotted for the involved courses and their defined objectives as per curriculum standards (see Fig. 9.3 for an example).

After implementing the I-STEM-PBL module, the teachers evaluate student learning. Following the I-STEM-PBL-ESD model, evaluation of learning should be in the form of MCU, CPAC, and KAB, which is in accordance with the definition of quality education (World Education Forum 2015: Para. 9). However, although the model suggests that the evaluation must focus on these three variables, the teachers are not limited to evaluating other student learning indicators.

On one hand, the connection among the real-world contextualized problems, I-STEM-PBL module utilization, and student learning in accordance with quality education definition, may lead to the promotion of ESD. On the other hand, the status of ESD will create feedback that would guide teachers in formulating problems, improving the implementation of I-STEM-PBL modules, and ensuring student learning.

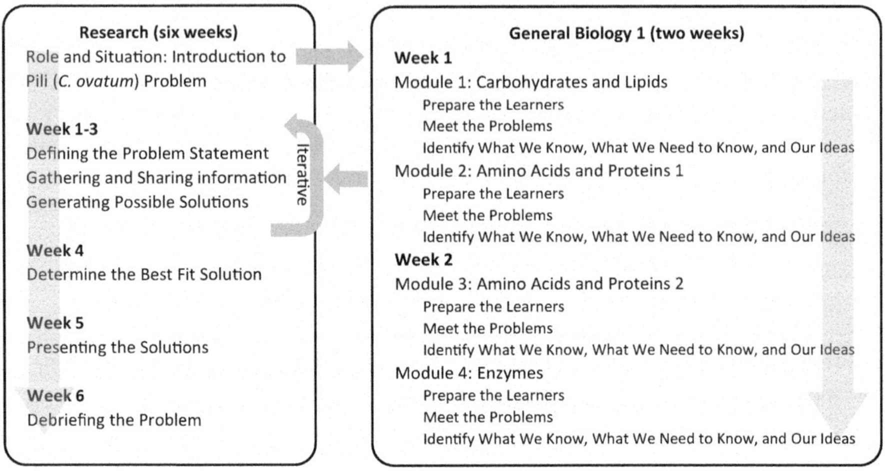

Fig. 9.3 Sample implementation plan of the I-STEM-PBL module in biological molecules and research courses for grades 10 and 11. *Note* This is a hypothetical implementation plan for the Philippine K-12 curriculum

9.3 Conclusion and Recommendation

The unprecedented risks and uncertainties have made amendments in education, emphasizing the preparation of the youth to face real-world problems. This resulted in the need for a proactive integration of SD in the learning and teaching processes through ESD. Innovations to implement ESD improve its capacity as the key enabler for SD and quality education, answering to the changing demands in education.

On this note, the researchers developed a model called the I-STEM-PBL-ESD instructional framework that aims to integrate classical and innovative techniques in learning and teaching to promote SD; respond to UNESCO's urgent appeal to scale up educational interventions that promote quality education; and inculcate in students of all ages the knowledge, skills, attitudes, and behaviors necessary to confront real-world challenges and advance SD. Understanding the processes involved in this framework may help teachers and instructional designers to develop an effective and efficient way of implementing learner-centered action-oriented transformative pedagogy, I-STEM approach, and PBL; and explore student learning in the areas such as conceptual understanding, creativity, knowledge, attitudes, and behaviors towards SD. In addition, this framework not only serves as a guide for designing effective educational modules, materials, and methods but also provides a research paradigm for examining the effects of the abovementioned variables on student learning and ESD.

Further studies and evaluation of the validity and robustness of the I-STEM-PBL-ESD framework for directing instructional designers and teachers towards developing quality and effective modules, materials, and methods, as well as the relationship and association between and among the significant variables within this framework, are needed. Because this study is limited to the variables listed, more research studies aimed at enhancing the I-STEM-PBL-ESD instructional framework and including other variables that are deemed important are recommended. For instance, researchers may investigate the effects of this framework in different learning modalities, namely: face-to-face, online distance learning, synchronous, asynchronous, or HyFlex classes, among others.

9.4 Note to Our Future Colleagues

In the year 2050, the world will undoubtedly be a vastly different place than it is today. As we have witnessed with the COVID-19 pandemic, unforeseen events can reshape our lives in profound ways. This raises the question: How can we prepare for the uncertainties of the future, and what role does ESD play in this endeavor?

The COVID-19 pandemic serves as a stark reminder of the unexpected challenges that can disrupt our lives and institutions. It brought education to a grinding halt, forcing a sudden shift to remote learning and laying bare the digital divide that persists in many parts of the world (Talidong and Toquero 2020; Williamson et al. 2020). While we cannot predict every crisis, we can equip ourselves with tools and mindsets that foster resilience and adaptability. ESD, in this context, emerges as a crucial instrument not only for achieving the 2030 SDGs but also for navigating the uncertainties of the future.

ESD goes beyond imparting knowledge; it instills a holistic understanding of the interconnectedness of economic, environmental, social, and political factors. In a world where challenges transcend borders and disciplines, this holistic perspective is invaluable. As we look ahead to 2050, ESD can serve as a compass for our leaders of tomorrow, guiding them to make informed decisions that balance the needs of people, planet, and prosperity.

One significant development on the educational landscape is the I-STEM-PBL-ESD model, which seeks to revolutionize the way we approach education. McComas and Burgin (2020) emphasize that when an insight fundamentally alters our scientific understanding, a revolution takes place. This model strives to do just that by redefining the relationships between Problem-Based Learning (PBL), STEM, and ESD. While debates about the precise definition of STEM education persist (Honey et al. 2014; Kelley and Knowles 2016; Koh 2018; McComas and Burgin 2020), the hope is that by 2050, a consensus will emerge, clarifying the role of STEM in achieving quality education.

In the interim, this research contributes to our evolving understanding of PBL, STEM, and ESD. It provides a framework for educators and policymakers to explore emerging ideas and adapt to the evolving landscape of education. As we prepare our students today, we must recognize that they are the leaders and problem solvers of tomorrow. Equipping them with the skills and knowledge fostered through ESD and innovative educational models like I-STEM-PBL-ESD is an investment in a more sustainable and resilient future.

Acknowledgements The authors would like to thank the Department of Science and Technology—Science Education Institute (DOST-SEI), Philippines through the Capacity Building Program in Science and Mathematics Education (CBPSME) for their support of this project.

References

Albanese MA, Mitchell S (1993) Problem-based learning: a review of literature on its outcomes and implementation issues. Acad Med 68:52–81. https://doi.org/10.1097/00001888-199301000-00012

Bakarman AA (2005) Attitude, skill, and knowledge (ASK) a new model for design education. In: Proceedings of the Canadian Engineering Education Association (CEEA). https://doi.org/10.24908/pceea.v0i0.3894

Barrows HS (1986) A taxonomy of problem-based learning methods. Med Educ 20(6):481–486. https://doi.org/10.1111/j.1365-2923.1986.tb01386.x

Barrows HS (1996) Problem-based learning in medicine and beyond: a brief overview. New Dir Teach Learn 28:3–12. https://doi.org/10.1002/tl.37219966804

Barrows HS, Myers AC (1993) Problem-based learning in secondary schools [unpublished monograph]. Problem-Based Learning Institute, Lanphier High School, and Southern Illinois University Medical School, Springfield, IL

Bryan LA, Moore TJ, Johnson CC, Roehring GH (2016) Integrated STEM education. In: Johnson CC, Peters-Burton EE, Moore TJ (eds) STEM Road Map: a framework for integrated STEM education. Routledge, New York, USA, pp 23–38

Bybee RW (2010) What Is STEM education? Science 329(5995):996–996. https://doi.org/10.1126/science.1194998

Bybee RW (2013) Case for stem education: challenges and opportunities. NSTA Press, Arlington, USA

Capraro RM, Capraro MM, Morgan JR (eds) (2013) STEM project-based learning: an integrated science, technology, engineering, and mathematics (STEM) approach. Springer Science & Business Media

Catelo SP, Jimenez SD (2016) An exploratory study of the economic potential of Philippine Pili pulp oil from waste pulp. J Econ, Manag, Agric Dev 2(1)

Department of Education (2016) K-12 science curriculum guide. DepEd. https://www.deped.gov.ph/wp-content/uploads/2019/01/Science-CG_with-tagged-sci-equipment_revised.pdf

Department of Education (2019a) Policy guidelines on the k to 12 basic education program (DepEd Order No. 021, s. 2019). Department of Education

Department of Education (2019b) Alternative delivery mode learning resource standards (ADMLR STANDARDS) [version 12]. Department of Education

Dochy F, Segers M, Van den Bossche P, Gijbels D (2003) Effects of problem-based learning: a meta-analysis. Learn Instr 13(5):533–568. https://doi.org/10.1016/S0959-4752(02)00025-7

Doyle T (2011) Learner-centered teaching: putting the research on learning into practice. Stylus Publishing, LLC

Duch BJ, Groh SE, Allen DE (eds) (2001) The power of problem-based learning. Stylus, Sterling, VA

Dugger W, Fellow J (2011) Evolution of STEM in the United States [Paper presentation]. In: 6th Biennial international conference on technology education research, Surfers Paradise. https://search.informit.org/doi/10.3316/informit.731822911107720

Ekantini A, Wilujeng I (2018) The development of science student worksheet based on education for environmental sustainable development to enhance scientific literacy. Univ J Educ Res 6(6):1339–1347. https://doi.org/10.13189/ujer.2018.060625

Elkington J (1994) Towards the sustainable corporation: win-win-win business strategies for sustainable development. Calif Manage Rev 36(2):90–100

El Sayary AMA, Forawi SA, Mansour N (2015) STEM education and problem-based learning. Routledge international handbook of research on teaching thinking. Routledge, pp 381–392

Evensen DH, Hmelo CE (eds) (2000) Problem-based learning: a research perspective on learning interactions. Lawrence Erlbaum Associates Publishers

Fakhrudin IA, Wicaksana EJ, Nastiti AR, Saljadziba E, Indriyanti NY (2021) Pre-service teachers' perspectives: STEM as a solution to promote education for sustainable development. J Phys: Conf Ser 1842(1):012082. https://doi.org/10.1088/1742-6596/1842/1/012082

Funa AA, Prudente MS (2021) Effectiveness of problem-based learning on secondary students' achievement in science: a meta-analysis. Int J Instr 14(4):69–84. https://doi.org/10.29333/iji.2021.1445a

Funa AA, Talaue FT (2021) Constructivist learning amid the COVID-19 pandemic: investigating students' perceptions of biology self-learning biology modules. Int J Learn, Teach, Educ Res 20(3). https://doi.org/10.26803/ijlter.20.3.15

Funa AA, Gabay RAE, Estonanto AJJ, Prudente MS (2022) Development and validation of online survey instrument on sustainable development for science teachers: focus on pili (canariumovatum). J Turk Sci Educ 19(2):559–576. https://doi.org/10.36681/tused.2022.137

Funa AA, Gabay RAE, Ibardaloza RT, Limjap AA (2022) Knowledge, attitudes, and behaviors of students and teachers towards education for sustainable development. Cakrawala Pendidikan: J Ilmiah Pendidikan 41(3):567–580. https://doi.org/10.21831/cp.v41i3.42407

Grant MM, Tamim SR (2019) PBL in K–12 education. In: Moallem M, Hung W, Dabbagh N (eds) The Wiley handbook of problem-based learning. Wiley Blackwell, pp 222–243

Hamilton JD (1976) The McMaster curriculum: a critique. BMJ 1(6019):1191–1196

Harmer AJ, Cates WM (2007) Designing for learner engagement in middle school science: technology, inquiry, and the hierarchies of engagement. Comput Schools 24(1–2):105–124. https://doi.org/10.1300/J025v24n01_08

Hewson PW, Hewson MGB (1984) The role of conceptual conflict in conceptual change and the design of science instruction. Instr Sci 13(1):1–13. https://doi.org/10.1007/BF00051837

Hmelo-Silver CE (2004) Problem-based learning: what and how do students learn? Educ Psychol Rev 16(3):235–266. https://doi.org/10.1023/B:EDPR.0000034022.16470.f3

Honey M, Pearson G, Schweingruber H (eds) (2014) STEM integration in K-12 education: status, prospects, and an agenda for research. The National Academy Press, Washington, DC

Hopkinson P, James P (2010) Practical pedagogy for embedding ESD in science, technology, engineering and mathematics curricula. Int J Sustain High Educ 11(4):365–379. https://doi.org/10.1108/14676371011077586

Hung W (2006) The 3C3R model: a conceptual framework for designing problems in PBL. Interdisc J Prob-Based Learn 1(1):6. https://doi.org/10.7771/1541-5015.1006

Hung W (2009) The 9-step problem design process for problem-based learning: application of the 3C3R model. Educ Res Rev 4(2):118–141. https://doi.org/10.1016/j.edurev.2008.12.001

Hung W (2019) Problem design in PBL. In: Moallem M, Hung W, Dabbagh N (eds) The Wiley handbook of problem-based learning. Wiley Blackwell, pp 249–272

Hung W, Moallem M, Dabbagh N (2019) Social foundations of problem-based learning. In: Moallem M, Hung W, Dabbagh N (eds) The Wiley handbook of problem-based learning. Wiley Blackwell, pp 51–79

International Technology and Engineering Educators Association (n.d.) Integrative stem education. Retrieved 26 Feb 2022, from https://www.iteea.org/IntegrativeSTEMEducation.aspx

Kelley TR, Knowles JG (2016) A conceptual framework for integrated STEM education. Int J STEM Educ 3(11):1–11. https://doi.org/10.1186/s40594-016-0046-z

Klarin T (2018) The concept of sustainable development: from its beginning to the contemporary issues. Zagreb Int Rev Econ Bus 21(1):67–94. https://doi.org/10.2478/zireb-2018-0005

Koh T (2018) Humanities at the heart of a holistic education in a tech-driven world. https://www.channelnewsasia.com/news/commentary/tommy-koh-humanities-education-fourth-industrial-revolution-10522334
Kolb DA (1984) Experiential learning: experience as the source of learning and development. Prentice-Hall
LaForce M, Noble E, Blackwell C (2017) Problem-based learning (pbl) and student interest in STEM careers: the roles of motivation and ability beliefs. Educ Sci 7(4):92. https://doi.org/10.3390/educsci7040092
Laurie R, Nonoyama-Tarumi Y, Mckeown R, Hopkins C (2016) Contributions of education for sustainable development (ESD) to quality education: a synthesis of research. J Educ Sustain Dev 10(2):226–242. https://doi.org/10.1177/0973408216661442
Leicht A, Combes B, Byun WJ, Agbedahin AV (2018) From agenda 21 to target 4.7: the development of education for sustainable development. In: Leicht A, Heiss J, Byun WJ (eds) Issues and trends in education for sustainable development. UNESCO, pp 25–38. https://en.unesco.org/sites/default/files/issues_0.pdf
Lopatto D (2004) Survey of undergraduate research experiences (SURE): first findings. Cell Biol Educ 3(4):270–277. https://doi.org/10.1187/cbe.04-07-0045
Lotz-Sisitka H, Wals AE, Kronlid D, Mcgarry D (2015) Transformative, transgressive social learning: rethinking higher education pedagogy in times of systemic global dysfunction. Curr Opin Environ Sustain 16:73–80. https://doi.org/10.1016/j.cosust.2015.07.018
Marrero MM, Gunning AE, Germain-Williams T (2014) What is STEM education? Glob Educ Rev 1(4):1–6. https://doi.org/10.1126/science.1194998
Mayer RE (1983) Schema theory: thinking as an effort after meaning. In: Thinking, problem solving cognition. Freeman, San Francisco
McComas WF, Burgin SR (2020) A critique of "stem" education. Sci Educ 29(4):805–829. https://doi.org/10.1007/s11191-020-00138-2
McKeown R, Hopkins C (2017) Quality education and education for sustainable development. Decade Prog Educ Sustain Dev 28
Michalos AC, Kahlke PM, Rempel K, Lounatvuori A, MacDiarmid A, Creech H, Buckler C (2017) Progress in measuring knowledge, attitudes and behaviours concerning sustainable development among tenth grade students in Manitoba. Development of quality of life theory and its instruments. Springer, Cham, pp 301–336
Millena CG, Sagum RS (2018) Philippine Pili (Canarium ovatum, Engl.) varieties as source of essential minerals and trace elements in human nutrition. J Food Compos Anal 69:53–61. https://doi.org/10.1016/j.jfca.2018.02.008
Miller AL (2014) A self-report measure of cognitive processes associated with creativity. Creat Res J 26(2):203–218. https://doi.org/10.1080/10400419.2014.901088
Morris TH (2020) Experiential learning–a systematic review and revision of Kolb's model. Interact Learn Environ 28(8):1064–1077. https://doi.org/10.1080/10494820.2019.1570279
Moust JHC, van Berkel HJM, Schmidt HG (2005) Signs of erosion: reflections on three decades of problem-based learning at Maastricht University. High Educ 50(4):665–683. https://doi.org/10.1007/s10734-004-6371-z
Neufeld VR, Barrows HS (1974) The "McMaster Philosophy": an approach to medical education. J Med Educ 49:1040–1050. https://doi.org/10.1097/00001888-197411000-00004
Olsson D, Gericke N, Rundgren SC (2015) The effect of implementation of education for sustainable development in Swedish compulsory schools—assessing pupils' sustainability consciousness. Environ Educ Res 22(2):176–202. https://doi.org/10.1080/13504622.2015.1005057
Piaget J (1929) The two directions of scientific thought. Arch Phys Nat Sci 11:145–162. https://doi.org/10.1093/oxfordhb/9780199734689.013.0035
Posner GJ, Strike KA, Hewson PW, Gertzog WA (1982) Accommodation of a scientific conception: toward a theory of conceptual change. Sci Educ 66(2):211–227. https://doi.org/10.1002/sce.3730660207

Rieckmann M (2018) Learning to transform the world: key competencies in education for sustainable development. In: Leicht A, Heiss J, Byun WJ (eds) Issues and trends in education for sustainable development. UNESCO, pp 39–59. https://en.unesco.org/sites/default/files/issues_0.pdf

Romito LM, Eckert GJ (2011) Relationship of biomedical science content acquisition performance to students' level of PBL group interaction: are students learning during PBL group? J Dent Educ 75(5):653–664. https://doi.org/10.1002/j.0022-0337.2011.75.5.tb05091.x

Roth KJ (1990) Developing meaningful conceptual understanding in science. Dimensions of thinking and cognitive instruction, 1393175

Russell SH, Hancock MP, McCullough J (2007) Benefits of undergraduate research experiences. Science 316:548–549

Sachs J (2015) The age of sustainable development. Columbia University Press, New York

Sanders M (2009) STEM, STEM education, STEMmania. Technol Teacher 68(4):20–26

Savery JR (2019) Comparative pedagogical models of problem-based learning. In: Moallem M, Hung W, Dabbagh N (eds) The Wiley handbook of problem-based learning. Wiley Blackwell, pp 25–50. https://doi.org/10.1002/9781119173243.ch4

Schmidt HG (1983) Problem-based learning: rationale and description. Med Educ 17(1):11–16. https://doi.org/10.1111/j.1365-2923.1983.tb01086.x

Schmidt HG (1993) Foundations of problem-based learning—some explanatory notes. Med Educ 27(5):422–432. https://doi.org/10.1111/j.1365-2923.1993.tb00296.x

Schmidt HG, Van der Molen HT, Te Winkel WW, Wijnen WH (2009) Constructivist, problem-based learning does work: a meta-analysis of curricular comparisons involving a single medical school. Educ Psychol 44(4):227–249. https://doi.org/10.1080/00461520903213592

Schmidt HG, Rotgans JI, Yew EHJ (2019) Cognitive constructivist foundations of problem-based learning. In: Moallem M, Hung W, Dabbagh N (eds) The Wiley handbook of problem-based learning. Wiley Blackwell, pp 25–50. https://doi.org/10.1002/9781119173243.ch2

Slavich GM, Zimbardo PG (2012) Transformational teaching: theoretical underpinnings, basic principles, and core methods. Educ Psychol Rev 24(4):569–608. https://doi.org/10.1007/s10648-012-9199-6

Sockalingam N, Schmidt HG (2011) Characteristics of problems for problem based learning: the students' perspective. Interdisciplinary Journal of Problem-Based Learning 5(1):6–33. https://doi.org/10.7771/1541-5015.1135

Sorsogon Pili Board (2018) Status of Pili industry in Sorsogon and Bicol. Province of Sorsogon Archives

Sorsogon Pili Roadmap (2018) Pili roadmap. Sorsogon State College and Office of Sen. Chiz Escudero. Sorsogon City, Philippines. Province of Sorsogon

Suh H, Han S (2019) Promoting sustainability in university classrooms using a STEM project with mathematical modeling. Sustainability 11(11):3080

Suškevičs M, Hahn T, Rodela R (2019) Process and contextual factors supporting action-oriented learning: a thematic synthesis of empirical literature in natural resource management. Soc Nat Resour 32(7):731–750. https://doi.org/10.1080/08941920.2019.1569287

Talidong KJB, Toquero CMD (2020) Philippine teachers' practices to deal with anxiety amid COVID-19. J Loss Trauma 25(6–7):573–579. https://doi.org/10.1080/15325024.2020.1759225

Tan AL, Teo TW, Choy BH, Ong YS (2019) The S-T-E-M quartet. Innov Educ. https://doi.org/10.1186/s42862-019-0005-x

Torp L, Sage S (2002) Problem-based learning for K–16 education. Association for supervision and curriculum development. Alexandria, VA

Torraco RJ (2005) Writing integrative literature reviews: guidelines and examples. Hum Resour Dev Rev 4(3):356–367. https://doi.org/10.1177/1534484305278283

United Nations Department of Economic and Social Affairs (2020) Education for sustainable development: a roadmap. UNESCO, France. https://unesdoc.unesco.org/ark:/48223/pf0000374802.locale=en

United Nations Division of Sustainable Development (1992) Agenda 21. UNCED, Rio de Janerio, Brazil. https://sustainabledevelopment.un.org/content/documents/Agenda21.pdf
United Nations Educational, Scientific and Cultural Organization (2011) ESD: education for sustainable development—building a better, fairer world for the 21st century. UNESCO. http://www.unesco.org/new/en/education/themes/leading-the-international-agenda/education-for-sustainable-development/
United Nations Educational, Scientific and Cultural Organization (2014) UNESCO roadmap for implementing the global action programme on education for sustainable development. UNESCO, France. https://sustainabledevelopment.un.org/content/documents/1674unescoroadmap.pdf
United Nations Educational, Scientific and Cultural Organization (2017) Education for sustainable development goals. learning objectives. UNESCO. http://unesdoc.unesco.org/images/0024/002474/247444e.pdf
United Nations Educational, Scientific and Cultural Organization (2018) Issues and trends in education for sustainable development. UNESCO, France. https://en.unesco.org/sites/default/files/issues_0.pdf
Vasquez JA, Snider C, Comer M, Bybee R (2013) Stem lesson essentials, grades 3–8: Integrating science, technology, engineering, and mathematics. Heinemann
Weimer M (2013) Learner-centered teaching: five key changes to practice, 2nd edn. Jossey-Bass
Widiyatmoko A, Shimizu K (2018) An overview of conceptual understanding in science education curriculum in Indonesia. J Phys: Conf Ser 983:012044. https://doi.org/10.1088/1742-6596/983/1/012044
Williamson B, Eynon R, Potter J (2020) Pandemic politics, pedagogies and practices: digital technologies and distance education during the coronavirus emergency. Learn, Media Technol 45(2):107–114. https://doi.org/10.1080/17439884.2020.1761641
World Commission on Environment and Development (1987) Report of the world commission on environment and development: our common future. WCED. http://www.un-documents.net/our-common-future.pdf
World Education Forum [WEF] (2015) World education forum 2015 final report. United Nations Educational, Scientific and Cultural Organization
Zguir MF, Dubis S, Koç M (2021) Embedding education for sustainable development (ESD) and SDGS values in curriculum: a comparative review on Qatar, Singapore and New Zealand. J Clean Prod 319:128534. https://doi.org/10.1016/j.jclepro.2021.128534

Part III
History, Philosophy and Sociology of Science/Engineering and Informal Learning

Chapter 10
Commentary for Part III: Looking Back and Looking On as Science/STEM Education Evolves

Timothy Ter Ming Tan

10.1 Historical Perspectives

The third and final part of this book examines three chapters on seemingly disparate threads, but they all ultimately provide a perspective on the historical development of learning in science and engineering under changing circumstances, albeit at different time scales. This look back is particularly apt as we attempt to chart our titulary "pathways through science education", which I consider as encompassing both *where we have been* to *where are we going*. Examination of the former can help inform the latter as well as provide the background to *how we got here* and *why we do what we do*. In this case, the context is the teaching and learning of science and engineering in both formal (practical work, apprenticeship) and informal approaches.

Chapter 11 by William Palmer is a historical treatise on the work of eighteenth and nineteenth century academics at Harvard University to develop and advance the teaching of chemistry through the use of practical work at the undergraduate level. Through this chronicle spanning more than a century, the origins and early evolution of laboratory practical work for chemistry teaching and by extension science teaching in general can be seen. In contrast, Chap. 12 by Grohman and colleagues describes the sudden disruption brought on by the Covid-19 pandemic in engineering research laboratories and the subsequent adaptations among graduate level students to continue with their research and learning. Even though it is an ethnographic study of just two research laboratories, there is an evocative and wry familiarity to the turmoil depicted and, in that sense, represents a microcosm of what happened within the wider academic community. These changes in daily life happened within a matter of days, but the adjustments in overall research enterprise quickly followed (see Radecki and Schonfeld 2020), while the effects on higher education and research

T. T. M. Tan (✉)
Natural Sciences and Science Education Academic Group, National Institute of Education, Nanyang Technological University, Singapore, Singapore
e-mail: timothy.tan@nie.edu.sg

Y. S. Ong et al. (eds.), *A Diversity of Pathways through Science Education*,
https://doi.org/10.1007/978-981-97-2607-3_10

have continued to reverberate in the years since (Abdrasheva et al. 2022). Chapter 13 is a review by Chng and Tan of the educational research from the first two decades of the twenty-first century on the informal learning of science at science centers and museums. In that period, a general trend towards increasing presentation of socio-scientific issues in tandem with traditional science content in exhibits was discerned, as was an emerging push towards greater inclusivity, especially of visitors from disadvantaged backgrounds. The work reviewed in this chapter covered the years leading up to but just before the pandemic struck, essentially summarizing the state of informal learning at science centers and museums in the pre-pandemic era and thus a potential point of comparison for any developments hence.

10.2 Changing Approaches to Science Learning

The sense of the "struggles" inherent in effecting and/or adapting to change also emerges from the three chapters. From the constant efforts to secure funding, laboratory space and resources by individual faculty members, some working in isolation, to advance the teaching of chemistry through practical work at a time when such practices were virtually unknown and untested, to the pandemic-induced mass upheaval and difficulties in conducting research and graduate training, the narratives in Chaps. 11 and 12 are all too recognizable to those in academia. A more subtle struggle is implied in Chap. 13, where museum educators, trying to update exhibits to match evolving trends and expectations of visitors, have to find a balance in their choices between traditional and more progressive approaches in exhibit design and style of presentation, and in navigating science centers and museums in their role within the overall milieu of science education. However, the struggles were perhaps an inevitable corollary to the need in each case to adapt to changing circumstances.

Palmer states in Chap. 11 that the introduction of practical work and hence the change in the teaching of chemistry was due to "[d]emographic, cultural, and technological influences", and this might quite aptly describe the underlying factors shaping the gradual shifts in science center and museum exhibit designs discussed in Chap. 13 as well. While these influences may provide the contextual circumstances that exert some pressure for change, I think it important to consider the intentionality for change as well. The Harvard professors described in Chap. 11 desired to improve chemistry education by shifting from the established teaching approach based on imparting of theoretical knowledge towards an approach emphasizing practical work in which students learned experientially by individually conducting laboratory bench work. This use of practical work is of course widely recognized in the modern era as an essential approach in the teaching and learning of science for a good number of reasons (see Millar 2004), some of which must have been the motivation behind the desire to change. More recently, parallels can be seen in the gradual push to increase opportunities for and the importance of informal learning approaches, and to integrate them with the formal (school-based) learning of science (Eshach 2007; Kim and Dopico 2016). Again, museum educators would be aware of the benefits of informal learning, as well as evidence for other best practices such as the use of socio-scientific approaches, and this may provide added impetus for shifts in emphasis, not only along the formal-informal axis, but also from traditional (i.e., dogmatic) to emergent (e.g., dialectic) approaches.

I was surprised to learn that in 1886, as Palmer reported in Chap. 11, Harvard University had instituted a pre-requisite requirement for school students wishing to be admitted to science courses, to have completed an extensive set of laboratory experiments while still in school. In doing so, they introduced the use of practical work in schools, a practice that has remained a fixture since. This had not been some gradual trickle-down of curriculum from higher education to secondary schools but a purposeful imposition with a lasting and pervasive impact on how school science has been taught. On the other hand, it is my contention that practical work in schools, and even that in higher education, has changed little since its introduction. Indeed, it would seem that at least some of the practical work activities conducted in schools today are the same as those original ones. More to the point, Chap. 11 should remind us that the described methodology of following a scripted protocol of actions and subsequently recording the observed and *expected* outcome of the "experiment" has largely remained the norm (Songer and Kali 2022). In other words, this is the "cookbook" approach that has been described as the teaching of science as a "rhetoric of conclusions" which tends to convey scientific knowledge as "empirical, literal, and irrevocable truths" (Schwab 1962). If "demographic, cultural, and technological influences" had been important drivers of the seminal change towards practical work, then surely these same factors in the circumstances of the present-day should be more than enough to kick-start updates to the way science practical work is conducted. I am personally in favor of the open-ended, integrative STEM, and epistemic approaches described elsewhere in this book.

10.3 The Importance of Embodied Learning, Engagement, and Interaction

Another area of thematic commonality between the three chapters in this part is the value placed on embodied learning and learning through interaction and collaboration. The benefits of practical work and other forms of bodily participation by the learner for learning, not just for skills-based tasks but also for conceptual and even language learning, has been widely discussed and affirmed (Macedonia 2019; Millar 2004). The individuals who brought practical work (conducted by individual students) into the science curriculum must have intuitively felt this benefit, at least initially, and perhaps its contributions to enhance learning and learner engagement would have soon become self-evident. The embodied cognition from such physical interaction with apparatus and materials is also echoed in the discussions of the importance of visitor-exhibit interactions in Chap. 13. In addition, visitor-visitor,

visitor-staff interactions, and interpersonal dynamics were also highlighted as significant aspects that can contribute to the overall informal learning experience. When such interpersonal interactions are disrupted (as detailed in Chap. 12) the opportunities and affordances for learning, as a socially mediated process, are adversely affected. It should be noted that even though the "practical work" of the research laboratory had also been severely curtailed, it was reported that learning, in other forms, resumed when social interactivity was re-established through online meetings and so on. This bodes well as we move into the post-pandemic era where such online interactions for learning have become much more widely accepted and practiced. Grohman and colleagues attributed the success in continued professionalization and learning within the research laboratories studied to emergent adaptations in, firstly, the ability of the graduate-level learners to self-regulate and exercise agency. They "remain[ed] motivated and empowered" to modify their research work and hence their own learning in light of the restrictions imposed on what had previously been routine laboratory work. Secondly, those who were supervising the members of the laboratories developed organic instructional methods for remote feedback and scaffolding, which was of particular importance because self-regulated learners seek and greatly benefit from such feedback from their instructors and peers. They concluded that priming learners for self-regulated learning and routine mentor–mentee and peer-peer interactions in both formal/academic as well as informal/social settings, even if conducted via online meetings, would be important ways to prepare them for and manage such disruptive circumstances in future.

10.4 Our Future History

Where then, will our present pathway or pathways lead us? In thinking about what, to our colleagues in 2050 will examine as history, I began to wonder what our colleagues from the eighteenth and nineteenth centuries might have said to us had they been asked. Even casting a look back a mere 30 years from this writing to the last decade of the twentieth century (roughly the same timespan as looking forward to 2050), what could colleagues then have accurately predicted of the educational landscape we now inhabit? Hence, instead of trying to predict our future directions, I will leave my thoughts framed as hope. It was noted in Chap. 13 that the papers on informal learning were primarily from developed, western economies, with few from the far more populated regions of the world. A similar point was also made by my colleague in his commentary to Part II. One can hope that researchers from more diverse economies, both developed and emerging, can step up to conduct and present rigorous research in their respective locales and by extension, contribute to the development of STEM education there. I would also like to hope that when we look back on this era, it will not be remembered for its pandemic-induced challenges, but that it will be celebrated instead as a time when the educational technology innovations and the wider acceptance of remote learning triggered by Covid-19 afforded a vigorous expansion and reinvention of pedagogical approaches that prepared learners well for the second quarter of the twenty-first century.

References

Abdrasheva D, Escribens M, Sabzalieva E, Vieira do Nascimento DM, Yerovi Verano CA (2022) Resuming or reforming? Tracking the global impact of the COVID-19 pandemic on higher education after two years of disruption. UNESCO. https://unesdoc.unesco.org/ark:/48223/pf0000381749

Eshach H (2007) Bridging in-school and out-of-school learning: formal, non-formal, and informal education. J Sci Educ Technol 16(2):171–190. https://doi.org/10.1007/s10956-006-9027-1

Kim M, Dopico E (2016) Science education through informal education. Cult Sci Edu 11:439–445. https://doi.org/10.1007/s11422-014-9639-3

Macedonia M (2019) Embodied learning: why at school the mind needs the body. Front Psychol 10:2098. https://doi.org/10.3389/fpsyg.2019.02098

Millar R (2004) The role of practical work in the teaching and learning of science. High school science laboratories: role and vision meeting. National Academy of Sciences, Washington, DC

Radecki J, Schonfeld RC (2020) The impacts of COVID-19 on the research enterprise: a landscape review. Ithaka S+R. https://doi.org/10.18665/sr.314247

Schwab JJ (1962) The teaching of science as enquiry. In: Schwab JJ, Brandwein PF (eds) The teaching of science. Harvard University Press, Cambridge, MA

Songer NB, Kali Y (2022) Science education and the learning sciences: a coevolutionary connection. In: The Cambridge handbook of the learning sciences, pp 486–503

Chapter 11
Changes in the Teaching of Practical Chemistry During the Nineteenth Century at American Universities

William Pitt Palmer

11.1 Introduction

This study covers the efforts of six chemists over the period of one hundred and twenty years, all working at Harvard University, Cambridge, Massachusetts, to improve the teaching of chemistry through practical work. The story that follows is the story of how scientists improved the teaching of science by devising a printed manual which each student owned and wrote the answers to the questions asked by the author based on the results of the practical experiments that the individual completed. This system was originally devised for university students but was eventually adopted by most secondary schools in the United States. Large numbers of laboratory manuals have been produced with the system continuing until the present time. The author of this study has a collection of early manuals complete with student answers making each manual unique. Table 11.1 showing the dates relating to the chemists and their work at Harvard follows:

11.2 Aaron Dexter

Aaron Dexter was born on 11th November 1750 in Malden, Massachusetts. He attended Harvard University obtaining an A.B. degree in 1776. He then started his career in medicine, apprenticed under Samuel Danforth who was a Boston physician and chemist (Miles 1976a, p. 119). He also had some experience as ship's surgeon during the Revolutionary War (Cohen 1950a, p. 36). He did not attend medical school but received an honorary MD from Harvard (Morrison 1937, p. 169). He also began a successful medical practice in Boston. At that time, the main professional use

W. P. Palmer (✉)
Faculty of Education, STEM, Curtin University, Perth, Australia
e-mail: bill_palmer15@hotmail.com

Y. S. Ong et al. (eds.), *A Diversity of Pathways through Science Education*,
https://doi.org/10.1007/978-981-97-2607-3_11

Table 11.1 Notable chemists from Harvard

Name	Years (birth and death)	Year graduated from Harvard	Years lecturing at Harvard	Erving Professor
Aaron Dexter	1750–1829	1776	1783–1816	Yes
John Gorham	1783–1829	1804	1809–1827	Yes 1816
John White Webster	1793–1850	1815	1824–1850	Yes 1827
Josiah Parsons Cooke	1827–1894	1848	1850–1894	Yes 1851
Francis Humphreys Storer	1832–1913	Assistant 1851–53. Graduated 1855	1870–1907	No. Agricultural Chemistry at Harvard 1870–1907
Charles William Eliot	1834–1926	1856	1858–1863	No. President of Harvard 1869–1909

for chemistry was in medicine. Prospective doctors needed an element of chemistry in their training and a few doctors expanded their knowledge of chemistry as a second string to their career. Dexter was one of these men. He kept up with the advances in chemistry by buying books from Europe. In 1783, he and two others (Dr John Warren and Dr Benjamin Waterhouse) were offered Professorships at Harvard Medical School as indicated by this newspaper report:

> They now inform the public, that they have chosen three professors, who have been approved by the Overseers, viz … and Dr Aaron Dexter for chymistry and the materia medica of Boston which town being so populous, and affording so large a theatre for practice, will give the professors the opportunity of introducing their pupils into the knowledge of many curious cases of physic and surgery. These gentlemen who have accepted the trust to which they have been elected, will soon be inducted into office. (Reporter 1783, p. 1)

Chemistry was only a minor portion of the medical course, so Dexter worked for four months of the year for the university and for eight months of the year in his private practice (Miles 1976a, p. 119). In 1790, Major William Erving, a good friend of Dexter's, died, leaving an endowment of £1000 to fund the Erving Professorship of Chemistry (Newell 1932, p. 689) from which Dexter benefitted as did Harvard. Dexter continued as Erving Professor of Chemistry and retired after serving for 33 years in 1816, becoming Professor Emeritus. He was succeeded by his former student John Gorham as he had planned; he died well-liked and respected on 28th February 1829 (Miles 1976a, p. 120).

A limited amount is known about Dexter's teaching over his long tenure. Firstly, it should be remembered that this was the time of the 'phlogiston debate'. Generally, the English-speaking world supported Joseph Priestley who believed that a curious substance called phlogiston existed and explained combustion whereas the Francophile chemists supported Lavoisier who explained combustion as being the chemical combination of a substance with oxygen. Dexter was unusual in that he supported Lavoisier and taught the 'new chemistry' and may indeed have been the first American to do so. Dexter used chemistry textbooks written by Professor Fourcroy, who favoured Lavoisier's theories, and in the surviving notebooks of his students, there appears to be no mention of phlogiston (Cohen 1950a, pp. 39–40). It was said of Dexter that 'He had not been withheld, by bigotry to early opinions, from adopting the current principles, which had been so splendidly announced from the French school.' (Jackson 1829, p. 7). Early on, Dexter had no chemical equipment though this situation gradually improved. An interesting feature of chemistry teaching at Harvard at this time was that Dr Benjamin Waterhouse also taught some chemistry and he did not accept the new French theories of Lavoisier. Dexter's own practical chemical skills were said to be modest (Miles 1976a, p. 120) as this quotation illustrates:

> "This experiment, gentlemen, is one of remarkable brilliancy. As I touch the powder you see before me with a drop of this fluid, it bursts into a sudden and brilliant flame" – which it most emphatically does not do as he makes the contact. "Gentlemen," he says, with a serious smile, "the experiment has failed, but the principle, gentlemen, the principle remains as firm as the everlasting hills." (Burrage 1920, p. 312)

A newspaper article provides evidence that he wrote about his analyses of potash for community benefit as he carefully described how potash could be manufactured.

> The potashes which I have examined, that have been condemned by the Assaymasters, I have found to contain principally common earth, which is undoubtedly the chief source of impurity in the potash of this country. … I have endeavoured to avoid prolixity and all chemical terms, as I wish to be understood by people concerned in this branch of business ...' (Reporter 1789 p. 2).

Dexter was involved in the analysis of potash for manufacturers, which would indicate that his practical chemical skills were greater than those usually credited to him.

11.3 John Gorham

John Gorham was born on 24th February 1783 to Stephen, a successful Boston merchant, and Molly (White) Gorham (Miles 1976b, p. 182). He obtained his A.B degree from Harvard in 1801 and his M.B from Harvard in 1804. He then went overseas to Paris, London and Edinburgh to study medicine but also returned with a great interest in chemistry. In 1808, he married Mary Warren who was the daughter of John Warren (Anon 1829, p. 107). Warren was a colleague at Harvard of Aaron Dexter and so Gorham became friendly with Dexter through him. Dexter had also been his professor. Gorham was appointed as an adjunct Professor of Chemistry at Harvard in 1809. In 1811, he obtained an M.D from Harvard (Elliott 1979, p. 106). After Dexter resigned in 1816, John Gorham became the Erving Professor of Chemistry. He held this position until 1827. Apart from his private practice and teaching chemistry at Harvard, he also taught geology from 1812 (when Waterhouse was dismissed) to 1824

(Frondel 1988, p. 1). Harvard, in 1812, expected that chemistry lecturers would pay for the chemical reagents that they used in their lectures whilst the university provided the equipment (Cohen 1950b, p. 88). At this time, doctors working in the Medical School also had a private practice which provided the major portion of their income. Gorham was a successful physician and the pressure of retaining his private practice and teaching chemistry at Harvard increased over time. He lived near his practice in Boston and when Harvard decided that professors should reside in Cambridge, his choice was to forgo his private practice or to resign his professorship. He resigned from the Erving professorship in 1827 and he died suddenly from pneumonia on 27th March 1829 (Reporter 1829, p. 3).

In 1805, whilst in London, Gorham and his friend Benjamin Silliman had lessons in experimental chemistry under Frederick Accum (Palmer 2011, pp. 554–558) as private pupils (Newell 1932, p. 690). He was a very active man in both medicine and chemistry and often combined the two skills with a compassionate and generous nature. For example, he traced the source of poison in a sample of sugar (Gorham 1817, pp. 228–235). He provided relief for poor insane patients in the Boston Almshouse using opium and camphor which was the best treatment at that time (Quen 1968, pp. 80–85). He was also an active member of the American Academy of Arts and Sciences, and many other associations (Reporter 1829, p. 3).

Between 1819 and 1820, he published a large two-volume work, 1200 pages long, entitled *The Elements of Chemical Science* (Gorham 1819). Gorham had originally intended this work to act as a textbook to accompany the lectures that he delivered to both the medical students and undergraduates that he was teaching. (Gorham 1819, p. ix). It was more successful than that because it was the first complete, systematic chemistry textbook written by an American and published in America (Newell 1932, p. 690). I B. Cohen described the textbook's influence as 'wide and profound' and stated that it presented 'a clear view of the state of chemistry' and 'it gave descriptions of many experiments.' (Cohen 1950b, pp. 83–85). The book evidently influenced Ralph Waldo Emerson, essayist, and poet, to take up the study of chemistry for a time (Browne 1932, p. 705). The book, *The apparatus of science at Harvard, 1765–1800* (Wheatland and Carson 1968, pp. 169–173) provides photographs of some of the apparatus that Gorham would have used and indicates that he used them experimentally. In 1817, he received a letter from the former president of the United States (1797–1801), John Adams encouraging his experiments in his pursuit of chemistry. Adams wrote:

> We are all chymists from our cradles. ... Chymists! pursue your experiments with indefatigable ardour and perseverance (quoted Weidlein 1928, p. 19)

Gorham's former student, John White Webster, joined Gorham at Harvard in 1824 and with Gorham's support succeeded him on his retirement (Miles 1976c, p. 499).

11.4 John White Webster

John White Webster was born on 20th May 1793 in Boston, Massachusetts. His parents were Redford Webster, a successful apothecary and Hannah (White) Webster (Elliott 1979, p. 269). He obtained his A.B at Harvard in 1811 and his M.D, also at Harvard, in 1815. He then travelled to London to study medicine at Guy's Hospital where the poet John Keats, a fellow student became a friend (Palmer 2013, p. 453). Webster was interested in geology and chemistry and whilst in Europe visited mineralogists in Britain and France, building up an extensive mineral collection (Miles 1976c, p. 499). On his return journey to America in 1817, he stopped for a few months in the Azores where he added to his geological specimens. He also met his future wife, Hannah Frederica Hickling, daughter of the American consul; they were married on 16th May 1818 and had four daughters (Susan, Marianne, Catherine and Harriet). He wrote a book on the geology of the Azores, which was published in 1821 (Webster 1821). Back in Boston, he started in medical practice with Doctor Gorham, but Webster's interest was in science rather than medicine. In 1818–1819, he gave a course of lectures on chemistry and mineralogy by subscription which were poorly attended with only four tickets sold (quoted Greene and Burke 1978, p. 77). Webster was appointed as a lecturer of chemistry, mineralogy and geology at Harvard in 1824, strongly recommended by Gorham. In 1826, Webster was promoted to adjunct Professor of Chemistry and to the position of Erving Professor Chemistry in 1827. Webster's reputation as a chemist mainly rests on his modifications of three existing European texts *A manual of chemistry*, (Webster and Brande 1826), *Elements of chemistry* (Webster and Fyfe 1827) and *Organic chemistry* (Webster et al. 1841) to suit the needs of American institutions. It is generally agreed that these books were a valuable contribution to the chemical literature of the time and showed Webster's chemistry as being up to date. Browne (1932, p. 706) saw *A manual of chemistry* as 'marking an important advance in our system of chemical education.' Webster was a joint editor of the *Boston journal of philosophy and the arts* with Daniel Treadwell and John Ware, but the journal only survived for three volumes. His writing sometimes reflected his teaching and research. For example, Werrett (2019, p. 43 and p. 49) shows that Webster in *A manual of chemistry* used unique equipment in the preparation of gases. Cohen (1950a, p. 43, Note 61) claimed that Webster was 'an important teacher, and his exemplary textbook [*A manual of chemistry*] was the vehicle for introducing American students to Faraday's discovery of the liquefaction and solidification of CO_2'. The opposite viewpoint was also expressed 'Once fairly in College, he [Cooke] distinguished himself in mathematics, but found little instruction in his favourite science' which was chemistry taught by Webster (Jackson et al., p. 514).

There are very different assessments of Webster's teaching competence in his twenty-six years at Harvard. For example: he was 'A kindly man liked by his pupils' (Miles 1976c, p. 500) or 'Not greatly distinguished as researcher or as teacher of chemistry …' (Elliott 1979, p. 269). He made a good start to his career but in his later years became overwhelmed with problems, particularly debt and university politics. His problems with debt seem have been self-inflicted and almost incomprehensible to an outsider. He came from a wealthy family and was left $50,000 by his father, Redford Webster, in 1833–1834 which solved his current financial problems. However, he made numerous unwise purchases including building a lavish house (Webster's Folly) which he was forced to sell; he then lived in rental accommodation (Chaney 2004, pp. 219–237). His university salary was $1200, and he was expected to pay for chemical reagents from that salary. In 1845, Webster claimed to have 'personally gone $5000 in debt to furnish his medical school laboratory' (Rossiter 1975, p. 69), whereas when Cooke started teaching in the laboratory in 1850, he found 'no apparatus worth mentioning' (Jackson et al. 1895, pp. 514–515).

In 1847, there were two contenders for the Rumford Chair, Eben Horsford, who Webster supported and Henry Rogers who Webster bitterly opposed. Two years earlier, Rogers had tried to have 'Webster ousted on charges of incompetence' (Rossiter 1975, p. 71). The fact that a charge of incompetence could be levelled against him with any credibility does indicate that the standards of his professional practice were slipping. His opposition to Rogers led to another attack on his position which occurred strangely enough on the day he was alleged to have committed murder. Harvard University appointed J. P. Cooke to teach the undergraduate chemistry course. Webster was to retain his position as Erving professor, but his salary would be reduced to $1000 per annum.

John White Webster's life has been the subject of perhaps more books and articles than any other American scientist because he was charged with murder, found guilty and hanged. The trial was controversial and raised important legal questions. Webster was accused of murdering George Parkman M.D., a very wealthy colleague who had loaned him money and who was hounding him for repayment. Webster had arranged to see Parkman in his laboratory on the afternoon that Parkman disappeared. Parts of a body that was found in the laboratory were identified as Parkman's and were said to have been skilfully dissected. However, some commentators doubt that Webster was the murderer, whilst others consider that Webster did not receive 'a fair trial' (Sullivan, quoted Miles 1976c, p. 500). It is widely accepted that the most likely murderer was John White Webster, and that the jury were correct, despite faults in the trial's procedure. There is a complete verbatim report of the trial (Webster et al. 1990).

Webster's teaching style would have been an improvement on that of his predecessor in terms of providing students with opportunities for practical work. Webster sponsored a chemistry club for students at Harvard called 'The Davy Club' where he was prepared to provide club members with bits of old apparatus and was always available for advice or information (Hale 1964). His methods were still the textbook, recitation, and lecture with the occasional spectacular demonstration, but there would have been no class practical work, particularly as it would have to have been financed from the lecturer's own pocket. His most famed experiment was the volcano experiment in which he added concentrated sulphuric acid to a mixture of sugar and potassium chlorate on a slab of soapstone. After the violent reaction, he evidently left the laboratory through a side door whilst his students had to exit via the windows (Jackson et al. 1895, p. 514).

Long (1901, pp. 360–372) surveyed the early teaching of chemistry in American medical schools and concluded that 'the introduction of even simple laboratory work is of comparatively recent date' with Harvard Medical School being the first in 1872. He was not convinced of the quality of the professors at medical schools and wrote that:

> The professor of chemistry was usually a physician who, as a rule, was not considered sufficiently strong to fill the chair of practice, obstetrics or surgery, but who might teach acceptably the less important branch of chemistry. (Long 1901, p. 360-372)

11.5 Josiah Parsons Cooke

Josiah Parsons Cooke was born in Boston, Massachusetts on 12th October 1827 to a wealthy family. His father was a well-known Boston lawyer and his mother, Mary (Pratt) Cooke died when he was only six (Siemiencow and Yusczuk 1976, p. 91). His secondary education was at the Boston Latin School. Whilst still at school, after being inspired by Benjamin Silliman Senior's public lectures, he set up a laboratory in a garden shed at his father's house, performing many experiments. He taught himself chemistry by repeating Silliman's demonstrations and by performing experiments included in Edward Turner's *Elements of Chemistry* (Contakes and Kyle 2011, p. 2). His favourite experiments were all with the newly discovered chemical processes which were friction matches, daguerreotype photography and guncotton (M.B. 1887, p. 377). His daguerreotype photographs are still preserved in Harvard University's Houghton Library (Rosen 1982, p. 525).

He attended Harvard, graduating in 1848. He was appointed as tutor in mathematics and instructor in chemistry in 1849 (Forbes 1971, p. 397). In 1851, after Webster's death, he was appointed Erving Professor of Chemistry and Minerology (Morrison 1937, p. 298). His appointment to his professorship was conditional on his purchasing at his own expense apparatus and equipment to run his courses (Siemiencow and Yusczuk 1976, p. 91). He obtained leave of absence for eight months to travel to Europe where he would be buying materials for teaching chemistry. He also attended lectures by Regnault and Dumas (Forbes 1971, p. 397). His health had been bad since he was a child, as he suffered from poor eyesight and a tremor of his hands. On his return to Cambridge, he took up his appointment in a difficult situation. He was convinced of the value of students doing practical chemistry themselves as the best means of teaching chemistry.

Justus von Liebig at the University of Giessen had devised the modern laboratory-oriented teaching method in about 1848. E. N. Horsford of the Lawrence Scientific School of Harvard had been a student of Liebig's (Forbes 1971, p. 397) and may have provided Cooke with details of Liebig's model, but Cooke's own experience was perhaps the major factor (Contakes and Kyle 2011, p. 18, note 2). He took action to provide more practical work for students step by step. In 1851, a lecture room was assigned to him in University Hall, and in the cellar beneath he fitted up a small research laboratory (M.B. 1887, p. 377) and used it to give practical instruction to select undergraduate students. Francis Humphreys Storer was employed as his assistant and Charles William Eliot was one of the undergraduates that he taught. Firstly, he organised chemistry courses for his students that used the traditional methods, textbooks, lectures, and recitations because he lacked laboratory space for practical work. He cleverly worked to increase laboratory facilities with support from the university, raising funds largely through his own efforts. He also set up an 'elective course in qualitative analysis, to be taught in the laboratory by the experimental method' (Jackson 1895, p. 178). He wrote his first paper on atomic weights in 1854 (Forbes 1971, p. 398). In 1858, a large building, Boylston Hall was opened; it was to be used for teaching chemistry but also housed an anatomical museum. This allowed the variety of chemical courses and his research to increase in quality and quantity. He then raised money using a 'matching funds' system where the university matched funds that he raised privately (Contakes and Kyle 2011, p. 3) to build a third floor for Boylston Hall. He then appealed for funds for the expansion of the university museum so it could hold the mineralogy collection which had been reorganised by retaining the services of Benjamin Silliman Senior (Forbes 1971, pp. 397–398). He used the space released to house a new organic chemistry department.

On 6th September 1861, Cooke married Mary Hinkley Huntingdon, daughter of Elisha Huntingdon and Hannah Hinkley. Mary was born on 3rd September 1833 and died 21st May 1911. They had no children (Longmore-Etheridge 2011).

He had built up a research department in chemistry with postgraduate students. These included Charles W. Eliot, Frank Austin Gooch, Henry Barker Hill, Charles Loring Jackson, and Theodore W. Richards (Contakes and Kyle 2011, p. 4). His own research covered chemistry and chemical education. In chemistry, his papers on determining the atomic weight of elements, were remarkable for their time and are mentioned by Partington (Partington 1965, pp. 878, 879 and 885). Cooke is the only one of the six chemists in this study to receive any attention in Partington's four volume *History of chemistry*. His doctoral student, Theodore W. Richards won the 1914 Nobel Prize in Chemistry (Contakes and Kyle 2011, pp. 1–2) for work initiated by Cooke who had provided this as a problem to Richards because of his failing eyesight (Ihde 1961, pp. 822–826).

It was in the field of chemical education that Cooke had his greatest success. He not only talked about the value of practical chemistry in teaching students, but he practised it. From the start of his career in 1850, he gave a course of general chemistry lectures (Chemistry A) to the entire freshman class in the College every year for forty-three years. The lectures were very popular with the students, providing brilliant demonstrations (Bliss 1940, p. 358) combined with a non-technical explanation of the nature of chemistry (Rosen 1982, p. 525). In his obituary, The New York Times stated:

> In order to interest the students, the old professor used to teach more through the eye than the ear and his dazzling experiments in electricity made before a crowded classroom, formed the chief delight of an otherwise dull freshman year (Reporter 1894, p. 13).

In 1871, he managed to negotiate the merger of the chemistry department of Harvard with that of the Lawrence School which overcame longstanding rivalries between the two departments (Rosen 1982, p. 525). By 1887, there were over three hundred students working in the new laboratories. There were thirteen distinct courses of instruction from all areas of the chemical sciences, and there were three professors, one instructor and three assistants employed in the teaching of chemistry (M. B. 1887, p. 377).

Cooke's writing was extensive; he wrote eight books, forty-one papers on chemistry and thirty-two papers on areas outside chemical research (Jackson 1895, pp. 180–183: Siemiencow and Yusczuk 1976, p. 92). He greatly influenced American chemical education through his books, his lectures, and his lecture demonstration techniques (Contakes and Kyle 2011, p. 4). He was a Unitarian (a Protestant Christian group). One of his most popular works was *Religion and chemistry* first published and later greatly revised (Cooke 1880). Twenty-seven English editions have been published between 1865 and 2019 (WorldCat). Cooke's views expressed in *Religion and Chemistry* follow in the tradition of William Paley's *Natural Theology* and William Prout in his Bridgewater Treatise *Chemistry, Meteorology, and Digestion Considered with Reference to Natural Theology*. Cooke spent the bulk of *Religion and Chemistry* trying to illustrate 'abundant evidence of design in the properties of the chemical elements alone' which related closely to his chemical research. (Browne 1932, p. 725; Contakes and Kyle 2011, p. 8). Cooke's textbook, *Elements of Chemical Physics* (Cooke 1860) which described the common physical methods employed by chemists was a very early book covering physical chemistry. Browne (1932, p. 724) commends it as being 'an epoch-making work, for it marked the dawn of the modern science of physical chemistry'. Cooke's text *First Principles of Chemical Philosophy* (Cooke 1868), provided a comprehensive description of chemical theory, supplemented with many problems and examples. His book, The *New Chemistry* (Cooke 1875), was an extremely popular work which contained a series of lectures he gave at the Lowell Institute in 1872; it explained the new approaches to chemistry developed since chemists accepted Avogadro's hypothesis following the first Karlsruhe International Chemistry Conference in 1860. In 1874, Professor Benjamin Silliman Jr. summarised Cooke's influence through his writing and personal example stating that:

> Professor Cooke has been largely instrumental in changing the older didactic methods of chemical instruction formerly in use, rendering them more exact and searching by a free use of the blackboard in the lecture room and laboratory (Silliman 1874, pp. 195-196).

Josiah Parsons Cooke passed away on 3rd September 1894 after being much weakened by a severe winter (Reporter 1894, p. 13).

11.6 Charles William Eliot, Francis Humphreys Storer and the Committee of Ten

Charles William Eliot was born in Boston, Massachusetts on 20th March 1834 to an influential family. His secondary education was at the Boston Latin School, and he obtained his B.A. from Harvard in 1853 and achieved his M.A. in 1856. As an undergraduate, he was the only student allowed in Cooke's research laboratory (Pratt 1976, p. 141). He wrote:

> I was the first student whom he [Cooke] admitted to the little laboratory in the basement of University Hall, and Professor Frank H. Storer and I were his first assistants, both at the Cambridge laboratory and at the laboratory in the Medical School. (Quoted, Davis 1929, p. 870)

Eliot was promoted from tutor in Mathematics in 1854 to an Assistant Professor in both Mathematics (1858–1861) and Chemistry (1858–1863). In chemistry he organised laboratory classes for undergraduates. From 1861, he oversaw the chemical laboratories at Lawrence Scientific School. At the end of his contract in 1863, he decided that he would pursue a career in teaching, leaving Harvard to devote two years to travel and study in Europe. He concentrated on finding out about the latest methods of teaching physics and chemistry and about the way in which technical education was organised. On his return in 1865 to the United States, he turned down a very generous commercial offer, instead taking the lowlier paid position of Professor of Analytical Chemistry and Metallurgy at the Massachusetts Institute of Technology where he taught with his old friend, Francis H. Storer (Pratt 1976, pp. 141–142).

Francis Humphreys Storer was born in Boston, Massachusetts on 27th March 1832 to David Humphreys Storer and Abby Jane (Brewer) Storer. His father, David, was an M.D. from Harvard who became a Professor of Obstetrics; he had a second career as a naturalist, specialising in ichthyology (Elliott 1979, p. 243). Francis Storer gained an interest in science from his father and attended the Lawrence Scientific School. He acted as assistant to Cooke from 1851 to 1853 whilst also teaching chemistry to a private class at Harvard Medical School. He obtained his B.S. degree from Harvard in 1855; he then spent two years in Europe studying with well-known chemists including Robert Bunsen in Heidelberg. He returned to the United States in 1857 and set up an industrial consulting firm. In 1865 he accepted an offer from the Massachusetts Institute of Technology (M.I.T.) to be Professor of General and Industrial Chemistry. He then worked with his friend, Charles William Eliot (Kopperl 1976, p. 461).

Eliot and Storer first needed to plan and equip the chemical laboratories in the Institute's new building to teach chemistry to all students by the laboratory method. The first M.I.T. catalogue stated that 'Practical instruction in chemical manipulations will also be given in the laboratory to every student' (Quoted Davis 1929, p. 874). It further stated that Chemistry courses at the Lowell Institute would be 'open to students of either sex, free of charge' (Quoted Davis 1929, p. 874) and would be given by Storer and Eliot. Robert H. Richards, a doctoral student of Storer's said of the M.I.T. course:

> He [Storer] and Professor Eliot were the pioneers in introducing the experimental method of giving instruction to classes in chemistry, and those who were ripe for it found in it the greatest inspiration (Richards 1915, p. 85)

They published a joint textbook *A Manual of Inorganic Chemistry Arranged to Facilitate the Experimental Demonstration of the Facts and Principles of the Science* (Eliot and Storer 1867) which ran to 672 pages and is available on Internet Archive; it was dedicated to Josiah Parsons Cooke. This text is usually seen as the first American laboratory manual, though its massive size hindered its practical use. Over the years this important textbook was modified several times with permission, firstly by William Ripley Nichols (Eliot et al. 1872). It was revised again by F. H. Storer and W. B Lindsay (Eliot et al. 1900). In each of the major revisions the text became more suited to secondary schools as well as colleges and universities.

After teaching for four years at M.I.T., on 1st July1869 Eliot resigned his position to become President of Harvard University at which point he broadened his research to the field of education more generally. Charles W. Eliot had married Ellen Derby Peabody in October 1858 and they had four children. Four days after the first news of his appointment as President of Harvard University, Eliot's wife, Ellen, died (Harvard Square Library, n.d.)

One of the first issues he concentrated upon was the future of the Bussey Institution, a constituent part of Harvard. Storer was appointed as the Professor of Agricultural Chemistry at the Bussey Institution in 1870; he married Eliot's sister, Catherine Atkins Eliot, in June 1871. He was promoted to become Dean of the Bussey Institution in 1871 and stayed in that position until his retirement in 1907 (Davis 1929, p. 878). Professor Storer died on 30th July 1914 with his wife predeceasing him (Richards 1915, p. 86).

Charles William Eliot now had strong views promoting educational reform which he made known in influential articles published in *Atlantic Monthly*. As President of Harvard, he immediately set about reforming the Medical Faculty with one of his major concerns being the standards of admission of the students. Eliot systematically devised ways of improving the quality of education delivered in each faculty. He also increased the student fees (Morrison 1937, p. 333) which allowed major building projects to progress (Harvard Square Library, n.d.). He asked Professor Cooke in 1886 to devise a course of at least sixty experiments which all science students should have completed prior to entering Harvard. Cooke wrote:

> A course of at least sixty experiments in 'general chemistry,' actually performed at school by the pupil.... The candidate will be required to pass both a written and a laboratory examination. The written examination will be directed to testing the candidate's knowledge of experiments and experimenting, as well as his knowledge of the principles and results of the respective sciences. The laboratory examination will be directed to testing his skill in experimenting. At the hour of the written examination the candidate will be required to hand in the original notebook in which he recorded the steps and results of the experiments which he performed at school, and this notebook must bear the indorsement of his teacher, certifying that the notes are a true record of the pupil's work. (Cooke 1886, p. 6)

This is significant as the university is demanding that secondary schools whose pupils wish to attend science courses at Harvard teach chemistry experimentally and certify that students have carried out experiments themselves. Eliot's presidency saw Harvard change from being a provincial college into a major American University famed for the quality of its research. He resigned on 19th May 1909 and in retirement edited the fifty volume Harvard Classics Series known as 'the five-foot shelf of books'. He died on 22nd August 1926.

In 1892, the Committee [of Ten] on Secondary School Studies was established at a meeting of the National Educational Association with the committee headed by Eliot, to review of American secondary school curricula. It was divided into nine subject-related subcommittees, one of which was the Physics, Astronomy, and Chemistry subcommittee (Kridel 2010, p. 125). Eliot was very influential in choosing the subcommittee members and he ensured that Ira Remsen (Palmer 2017, p. 6) was chair of the physical science subcommittee. This subcommittee made twenty-two recommendations to the full Committee of Ten. These included strong recommendations about laboratory work such as the recommendation that one-half of the time devoted to science subjects should be given to laboratory work and that careful notebook records of the laboratory work in both physics and chemistry should be kept by the student at the time of the experiment. (Sheppard and Horowitz 2006, pp. 569–570). Thus, the influence of Eliot and Remsen, emphasising laboratory-based university research was translated into the emphasis of practical science work in the secondary school. As many of the committee had written practical laboratory chemical manuals, a measure of self-interest could be involved, but the main aim was the improvement of chemistry teaching. The similarity between Cooke's sixty experiments (Cooke 1886, p. 6) and the Committee of Ten's proposals should be noted.

11.7 Conclusion

The story told is one of several that exist tracing the origins of student based practical work in America. According to Gamito-Marques's assertion, telling the story of the growth of practical instruction in chemistry and making it better known, should further improve the learning of chemistry. This account has emphasised the major part played by Josiah Parsons Cooke and the significant role played by Charles W. Eliot and Francis H. Storer in encouraging student practical work.

Note to our future colleagues

Dear colleagues,

In 2021, my paper looked back to view the progress in the teaching of chemistry by individualising student practical experience between 1780 and 1900. The claim of the paper is that the telling of this story also contributes to the student understanding of chemistry. The story also illustrates the acceleration of new knowledge in the ways of learning of chemistry which was accompanied by the acceleration of knowledge of chemistry itself.

Methods of learning will have changed by 2050, and one might anticipate a greater role for artificial intelligence, yet without massive societal changes it is difficult to believe that the necessity for individual human learning will have disappeared. The story of the early history of individualised instruction in chemistry will thus remain relevant.

The content of chemical knowledge has increased massively since 1780. Early in the nineteenth century, chemistry gradually came to be recognised as a subject separate from natural philosophy (Physics). By the end of the nineteenth century the subject of chemistry had separated into three sub-divisions which are inorganic chemistry, organic chemistry, and physical chemistry. These general divisions of chemistry remain useful today, whilst further new sub-divisions occurred as chemical knowledge has continued to expand. The nineteenth century saw the discovery of previously unknown elements with only thirty-three elements being known in 1800. By 1900 there were eighty-three known elements and there are now one hundred and eighteen known elements including those elements that have been produced artificially. The nineteenth century also saw the rise of spectroscopy which allowed chemists to identify elements from the spectra that they produced; this allowed chemists to identify the elements in the stars expanding the earth's chemical knowledge as being applicable to the whole universe. The twentieth century allowed the composition of atoms to be explored, finding out that atoms are made of smaller particles and investigating the properties of these particles.

By 2050, chemistry itself will have changed massively with the amount of chemical knowledge doubling every seven years, yet the laws which govern the ways in which substances interact seem unlikely to change. 'Dark matter' and 'dark energy' currently appear inexplicable and perhaps these features of distant worlds may be better understood in 2050. In 2021, many countries agreed to take action to limit the amount of carbon dioxide entering the atmosphere from the burning of fossil fuels. If the limit is achieved by 2050, you will be living a full and enjoyable life. If the target set has not been achieved, then you will now be suffering from the effects of the 'Greenhouse effect' and life on earth will have become very difficult. From the year 2021, I wish you every success in your new understandings of chemistry and apologise if we have left you an impossible climate situation through our own inaction.

References

Anon (1829) Death of Dr. Gorham. Boston Med Surgical J 2:107–109

Bliss AD (1940) Men and machines in early Harvard science. J Chem Educ 17(8):353

Browne CA (1932) The history of chemical education in America between the years 1820 and 1870. J Chem Educ 9(4):696–728

Burrage WL (1920) Dexter, Aaron (1750–1829). In: American medical biographies Kelly HA, Burrage WL (eds). The Norman Remington Company, Baltimore, p 312.

Chaney KE (2004) The Parkman Murder of 1849: The Commonwealth v. John White Webster. In: Bailey FY, Chermak S (eds) Famous American crimes and trials, vol 1: 1607–1859. Praeger, New York, pp 219–237

Cohen IB (1950a). The beginning of chemical instruction in America: a brief account of the teaching of chemistry at Harvard prior to 1800. In: Chymia: annual studies in the history of chemistry, vol 3, pp 17–44

Cohen IB (1950b) Some early tools of American science: an account of the early scientific instruments and mineralogical and biological collections in Harvard University. Russell and Russell, New York

Contakes SM, Kyle C (2011) Josiah Parsons Cooke Jr.: Epistemology in the service of science, pedagogy, and natural theology. Hyle: Int J Philos Chem 17(1):1–23

Cooke JP (1860) Elements of chemical physics. Little & Brown, John Allyn, Boston

Cooke JP (1868) First principles of chemical philosophy. Sever & Francis, Boston and Cambridge

Cooke JP (1875) The new chemistry. D. Appleton, New York

Cooke JP (1880) Religion and chemistry: a restatement of an old argument. C. Scribner's Sons, New York

Cooke JP (1886). The new requisitions for admission at Harvard College. Popular Science Monthly, vol 30, pp 1–11

Davis TL (1929. Eliot and Storer, pioneers in the laboratory teaching of chemistry. J Chem Educ 6(5):868–879

Eliot CW, Storer FH (1867) Manual of inorganic chemistry: arranged to facilitate the experimental demonstration of the facts and principles of the science. Boston. https://archive.org/details/manualofinrogani00eliorich/page/n661/mode/2up

Eliot CW, Nichols WMR, Storer FH (1872) An elementary manual of chemistry: abridged from Eliot and Storer's manual. Ivison, Blakeman, Taylor & Co, New York

Eliot CW, Storer FH, Lindsay WB, Nichols WR (1900) The compendious manual of qualitative chemical analysis of C.W. Eliot and F.H. Storer. Van Nostrand, New York

Elliott CA (1979) Biographies. In: Biographical dictionary of American science: the seventeenth through the nineteenth centuries. Westport, Connecticut, and London: Greenwood Press

Forbes G (1971) Cooke, Josiah Parsons Jr. In: Gillispie CC (ed) Dictionary of scientific biography (Caba-Deche), vol 3. American Council of Learned Societies and Charles Scribner's Sons, Publishers, New York, pp 397–399

Frondel C (1988) The geological sciences at Harvard University from 1788 to 1850. Earth Sciences History 7(1):1–22

Gamito-Marques D (2020) In praise of a historical story-telling approach in science education. Isis 111(3):582–587

Gorham J (1817) Chemical examination of a quantity of sugar, supposed to have been intentionally poisoned. N Engl J Med 6:228–235

Gorham J (1819) The elements of chemical science. Cummings and Hilliard, Boston. https://archive.org/details/2555027RX1.nlm.nih.gov/page/n37/mode/2up

Greene JC, Burke JG (1978) The science of minerals in the age of Jefferson, vol 68, pt 4. Transactions of the American Philosophical Society, Philadelphia

Hale EE (1964) A new England boyhood. Little, Brown, Boston (Chapter 10)

Harvard Square Library (n.d.) Eliot, Charles W. (1834–1926). Retrieved 9 May 2021 from https://www.harvardsquarelibrary.org/biographies/charles-w-eliot-harvard-university-president/

Ihde A (1961) American chemists at the century's turn. In: Farber E (ed) Great chemists. Interscience Publishers, New York and London, pp 822–826

Jackson CL (1895) Biographical memoir of Josiah Parsons Cooke. (1827–1894). Retrieved 25 April 2021 from http://www.nasonline.org/publications/biographical-memoirs/memoir-pdfs/cooke-josiah-p.pdf

Jackson J (1829) An address delivered at the funeral of John Gorham MD on the 31st day of March 1829, by the appointment of the Boston Medical Association, and published at their request. Boston Daily Advertiser, Boston. Retrieved 20 April 2021 from http://resource.nlm.nih.gov/101190019

Jackson CL, Hill HB, Lowell A, Storer FH, Eliot CW (1895) Proceedings [with a commemoration of Josiah Parsons Cooke]. Proc Am Acad Arts Sci 30:503–552

Kopperl SJ (1976) Francis Humphreys Storer (1832–1914). In: Miles WD (ed) American chemists and chemical engineers. American Chemical Society, Washington, DC, pp 461–462
Kridel C (ed) (2010) Encyclopedia of curriculum studies, vol 1. Sage Publishers, Los Angeles, p 125
Long JH (1901) Some points in the early history and present condition of the teaching of chemistry in the medical schools of the United States. Science (New Ser) 14(349):360–372
Longmore-Etheridge A (2011) Josiah Parsons Cooke, Harvard Professor, 1871, Albumen Cabinet Card. Flickr. Retrieved Sep 21, 2022 from https://www.flickr.com/photos/60861613@N00/5540694319/
M. B. (1887). Josiah Parsons Cooke. Sci Am 57(24):377 (10th December)
Miles WD (1976a) Aaron Dexter (1750–1829). In: Miles WD (ed) American chemists and chemical engineers. American Chemical Society, Washington, DC, pp 119–120
Miles WD (1976b) John Gorham (1783–1829). In: Miles WD (ed) American chemists and chemical engineers. American Chemical Society, Washington, DC, p 182
Miles WD (1976c) John White Webster (1793–1850). In: Miles WD (ed) American chemists and chemical engineers. American Chemical Society, Washington, DC, pp 499–500.
Morrison SE (1937) Three centuries of Harvard 1636–1936. Harvard University Press, Cambridge, Massachusetts
Newell LC (1932) Chemical education in America from the earliest days to 1820. J Chem Educ 9(4):677–695
Palmer WP (2011) Frederick Accum (1769–1838) and the application of chemistry to social problems. In: Sereoglou F, Koulountzos V, Siatras A (eds) Science and culture: promise, challenge and demand (book of proceedings). Epikentro Publications, Athens, pp 554–558
Palmer WP (2013) Professor John White Webster, science educator and convicted murderer. Chem: Bulgarian J Sci Educ 22:452–471
Palmer WP (2017) Ira Remsen and the sweet bitterness of chemistry, Chemical Education Program (Friday 28th July) at the RACI National Centenary Conference. Melbourne Conference Centre, Melbourne, Victoria, Australia
Partington JR (1965) (Reprint 1998) A history of chemistry, vol 4. Mansfield Centre, CT: Martino Fine Books
Pratt HT (1976) Charles William Eliot (1834–1926) In: Miles WD (ed) American chemists and chemical engineers. American Chemical Society, Washington, DC pp 141–143
Quen JM (1968) Early nineteenth-century observations on the insane in the Boston Almshouse. J History Med Allied Sci (jan) 23(1):80–85
Reporter (1783) University in Cambridge. Vermont J Windsor Vermont 1
Reporter (1789) Manufacturers of potash. Freeman's J North-Am Intell Philadelphia, Pennsylvania 2
Reporter (1829) Death of Dr. Gorman. J Times (Bennington, Vermont) 3
Reporter (1894) Prof. Cooke mourned at Harvard. New York Times 13
Richards. (1915) Francis Humphreys Storer. Science 41(1046):85–86. https://doi.org/10.1126/science.41.1046.85
Rosen S (1982) Josiah Parsons Cooke of Harvard. J Chem Educ 59(6):525
Rossiter MW (1975) Emergence of agricultural science—Justus Liebig and the Americans 1840–1880. Yale University Press, New Haven and London
Sheppard K, Horowitz G (2006) From Justus von Liebig to Charles W. Eliot: the establishment of laboratory work in U.S. high schools and colleges. J Chem Educ 83(4):566–570
Siemiencow G, Yusczuk R (1976) Josiah Parsons Cooke (1827–1894). In: Miles WD (eds) American chemists and chemical engineers. American Chemical Society, Washington, DC, pp 91–92
Silliman B Jr (1874) An essay on American contributions to chemistry. Am Chem 5(6):195–210
Webster JW (1821) A description of the island of St. Michael: Comprising an account of its geological structure. Published by R.P. & C. Williams, Boston

Webster JW, Brande WT (1826) A manual of chemistry, on the basis of Professor Brande's: containing the principal facts of the science, arranged in the order in which they are discussed and illustrated in the lectures at Harvard University, ... Richardson and Lord, Boston

Webster JW, Fyfe A (1827) Elements of chemistry. Richardson and Lord, Boston

Webster JW, Liebig J, Playfair LP (1841) Organic chemistry in its applications to agriculture and physiology. J. Owen, Cambridge

Webster JW, Stone JW, Massachusetts (1990) Report of the trial of Professor John W. Webster. Notable Trials Library, Birmingham, Alabama

Weidlein ER (1928) Industrial changes due to chemistry. In: The annals of the American academy of political and social science, September, vol 139, Stabilization of commodity prices, pp 15–33 (p 19)

Werrett S (2019) Household oeconomy and chemical inquiry. In: Thrifty science: making the most of materials in the history of experiment. University of Chicago Press, Chicago. https://discovery.ucl.ac.uk/id/eprint/10061612/1/[9789004325562%20-%20Compound%20Histories]%201_Household%20Oeconomy%20and%20Chemical%20Inquiry.pdf

Wheatland DP, Carson B (1968) The apparatus of science at Harvard, 1765–1800. Harvard University, Cambridge, Mass.

Chapter 12
The Formation of Engineers in Research Labs during the COVID-19 Crisis

Magdalena G. Grohman, Eun Ah Lee, Nicholas Gans, Ann Majewicz Fey, and Matthew J. Brown

12.1 Introduction

In recent years, pedagogical approaches to learning in higher education have gradually shifted away from didactic, lecture-based instruction to *pedagogies of engagement* that focus on turning students into lifelong, creative, and innovative learners. The pedagogies of engagement, which have their roots in Deweyan progressive education movement (lining experience with reflection and understanding with doing; Dewey 1933, 1938), embrace students' engagement and direct their experiences with methods and processes of inquiry and experiential learning (Edgerton 1997; Thompson 2014). The pedagogies of engagement no longer view educators as dispensers of knowledge and information, but as facilitators of learning who help

---

Dr. Magdalena G. Grohman is now Clinical Associate Professor at the Department of Multidisciplinary Innovation, Collage of Applied Collaborative Studies, University of North Texas, Denton, TX.

Dr. Eun Ah Lee is now Library Instructor and Research Specialist at the Public Library in Plano, Texas.

M. G. Grohman (✉) · E. A. Lee
Center for Values in Medicine, Science, and Technology, University of Texas, Dallas, TX, USA
e-mail: magdalena.grohman@unt.edu

N. Gans
Automation and Intelligent System Division, The University of Texas, Arlington, TX, USA

A. M. Fey
Department of Mechanical Engineering, The University of Texas, Austin, TX, USA

M. J. Brown
Center for Dewey Studies, School of History and Philosophy, Southern Illinois University, Carbondale, IL, USA

Y. S. Ong et al. (eds.), *A Diversity of Pathways through Science Education*,
https://doi.org/10.1007/978-981-97-2607-3_12

students stay engaged and hone their self-regulation, deliberate practice, and deep learning skills (Hodges 2015; Lang 2016).

It is within the pedagogies of engagement that we can see an important role that authentic research experiences play in high-quality engineering education and STEM education more broadly. Real-world research settings have been considered as a model of how students should learn, and engineering pedagogy experts recommend courses that imitate research laboratories or place students in working laboratories to experience authentic research. However, the recommendations for such research experiences have not received the rigorous examination they deserve, given their increasing prominence in engineering pedagogy.

Most of the previous studies on students' research experience focus on undergraduate students who choose to participate in actual laboratory research (Undergraduate Research Experience or URE) or courses with designed research experience (Course-based Undergraduate Research Experiences or CURE) and rarely include graduate students. According to Linn et al. (2015; see also Madan and Teitge 2013) those studies focus on learning outcomes by identifying the impact and benefits of URE and CURE rather than focusing on the learning process. Among the benefits and impact, those most often studied are retention and identity, understanding the nature of science and scientific practice, level of conceptual knowledge, and the impact of mentoring on students' future choices and attitudes (Frantz et al. 2006).

The majority of the reported studies are based on the pretest—treatment—posttest design, reflecting the comparisons of outcomes before and after participating in URE or CURE (see Linn et al. 2015). Additionally, assessments vary from retrospective surveys (~50% of 60 cases in Linn et al. 2015) through mixed methodologies (surveys and interviews) to interviews, study groups, and focus groups (Murdoch-Eaton et al. 2010). Only one study out of 60 has a core methodology that includes a qualitative method of direct observation (Linn et al. 2015). In CURE, the learning goal and learning outcomes are mostly well-defined, therefore surveys and knowledge/skills assessment may be sufficient to measure them. In URE, however, measuring outcomes of authentic research experiences depends on the specification of learning goals in a given laboratory context. For some of these goals, especially those pertaining to enculturation into scientific practice, cognitive-ethnographic methods of study are more justifiable. However, most of the existing studies, even those lending themselves to an ethnographic approach, often rely on surveys, test-based assessments, and grade-point averages (Slovacek et al. 2012).

As a consequence, the existing studies recognize what has been experienced and learned, but they fail to identify specific processes of learning. Moreover, studies based purely on surveys and test-based assessments offer only limited perspectives on enculturation processes into scientific communities that take place in research laboratories, especially if we consider the population of students who transition from undergraduate to graduate programs. On the other hand, the ethnographic approach to studying the impact and benefits of authentic research experiences creates an opportunity to do more direct and in-depth analysis of learning that takes place in the labs (cf. Linn et al. 2015).

It is also important to understand how research pedagogy is affected by long-term societal crises. The COVID-19 pandemic was severely disruptive to institutions of higher education. In the US, almost all universities halted on-campus instruction and research in the spring of 2020. According to the College Crisis Initiative at Davidson College, in fall 2021 only 27% of universities operated primarily in person, while 44% were fully online even through spring 2021. Online operation included every arm of colleges and universities, administration, education, and research. However, as education at universities transitioned to remote and hybrid models of instruction, faculty, staff, and students engaged in research could not return to their laboratory spaces to do research activities for over a year. In order to salvage any possibility of continuing their research and education mission, research laboratories had to adapt to the COVID-19 safety and risk mitigation requirements implemented by the universities, which stipulated that some of the lab research activities, like formal and informal lab meetings, had to be performed through online conferencing platforms. If the work absolutely demanded individuals to be in the lab, they had to take turns working alone, without their colleagues and mentors readily available to guide their research experience and learning.

As mentioned earlier, professional formation and training of future scientists and engineers are not limited to classrooms; rather, they take place while participating in authentic research experiences in working university research labs, and many, if not all, authentic research experiences depend on social interactions with peers, colleagues, and mentors. COVID-19 impacted such social interactions in the research lab and inevitably altered the process of the training and formation of research engineers. As researchers studying learning and formation of engineers in laboratory settings, we began to look at how this shift in response to COVID-19 changed the nature of these processes.

Science educators may wonder how our observations in engineering laboratories are related to science education at universities and in K-12 classrooms. The similarities in the scientific inquiry across the domains of science, engineering, and technology have been long recognized by leaders in STEM education, and systematic effort has been made to articulate them in science curricula. The Next Generation Science Standards (NGSS lead states, 2013), one of such efforts, emphasizes not only core disciplinary ideas but also science and engineering practices (e.g. planning and carrying out investigation), and cross-cutting concepts (e.g. stability and change or cause and effect). According to NGSS vision statement, emphasizing similarities between science and engineering will inspire students to pursue careers in science and engineering (National Research Council 2011). Our observations show that scientists and engineers alike are engaged in producing knowledge through similar practices of scientific inquiry across research laboratories in the institutions of higher education. As we have observed, engineers do not only engage in designing the solution to a current problem using existing technology. Often their research involves addressing complex challenges that require application of scientific inquiry practices such as formulating problems, planning, and carrying out investigations, designing experiments, analyzing data, and disseminating results. We believe that our observations in engineering research laboratories provide valuable examples of how scientific inquiry and engineering design practices unfold in real life and we hope that science educators and science students will find them inspiring.

In this chapter, we share our experiences with attempts to capture learning processes in two engineering labs during the COVID-19 pandemic. Two questions guided our inquiry: How does learning actually happen in authentic laboratory experiences? What are the effects of the COVID-19 pandemic on the learning process and authentic research experiences in the engineering labs? In the subsequent sections, we will briefly focus on the accounts of learning in the laboratory and on the processes closely tied to learning within pedagogies of engagement and authentic research experiences: self-regulation and feedback; we will then proceed to focus on the two engineering labs we observed and provide analysis of the kind of learning that unfolded before and during COVID-19.

12.1.1 Learning in the Laboratory

Our research project fits within the larger body of research in *laboratory studies* inaugurated by the work of Latour and Woolgar in *Laboratory Life* (1979/1986; see also Knorr Cetina 1995, 1999). This work adopts the attitude of "the anthropologist in the laboratory," that is, it presents the perspective of an ethnographic researcher who is totally naïve to the concepts and practices of science. Latour and Woolgar use this framing to examine the cultural practices of the laboratory, focusing on topics like the aims of the laboratory (to produce papers), the nature of laboratory equipment (inscription devices), the organization of the physical space, material resources, human activities within the laboratory, and most famously, the construction of scientific facts. The anthropological framing allows them to analyze these topics without short-circuiting their observations with descriptions based on the internal perspectives of scientists while ignoring social influences on the development of scientific knowledge. While they relax this pretense as the book goes on, in the second edition postscript, Latour and Woolgar famously eschew "far-fetched cognitive explanations over simpler social ones" (p. 274) and even propose a (probably tongue-in-cheek) "ten-year moratorium on cognitive explanations of science" (p. 280). From our perspective, the problematic assumption here is that social and cognitive explanations form a *dichotomy*, instead of considering the possibility that *social activities* realize or instantiate *cognitive processes* (Giere and Moffatt 2003).

More directly, our approach to studying authentic research experiences in engineering labs was inspired by the work of the Cognition and Learning in Interdisciplinary Cultures (CLIC) group at Georgia Tech, led by Nancy Nersessian and Wendy Newstetter. In their studies, CLIC uses a combination of ethnographic methods and cognitive-historical analysis of activities in a biomedical engineering research lab (Nersessian and Newstetter 2013; Newstetter et al. 2002, 2004). This particular combination of methodological tools allowed CLIC to demonstrate that a bioengineering lab is a constantly evolving distributed cognitive system, in which research and learning are intertwined and cannot be separated. Specifically, learning occurs as the interactions between lab members or between lab members and artifacts evolve, and this very same process contributes to research, resulting in conceptual and methodological knowledge distributed among lab members and artifacts (Nersessian 2006; 2009; 2019). In other words, in line with the critique of Latour and Woolgar above, the knowledge and cognition of the laboratory are realized through the social organization of the laboratory. When taking a close look at the interconnection and distribution of knowledge and cognition and the underlying social organization of the laboratory, Newstetter et al. (2004) observed that engineering research labs have a non-hierarchical organization in which no one person is *the* expert, that an interactional structure encourages participation and motivates participants, and that a strong social support system facilitates resiliency. Moreover, they also identified that the non-hierarchical organization of the engineering lab creates an environment conducive to *agentive* learning, in which learners interact with other agents of learning, including people and artifacts; are active, empowered, and motivated to seek learning opportunities and activities. In short, they see engineering research laboratories as sites of "robust and speedier learning."

Three implications from CLIC research informed our approach to studying the processes of the professional formation of engineers in a research laboratory. First the idea that an engineering research lab is an evolving cognitive distributed system led us to attempt to document short-term learning occurring at a given time and long-term learning unfolding over longer periods, and to analyze the distributed learning at the level of an entire lab, as well as at the level of an individual member. Second, the characteristics of an agentive learning environment gave us a framework to analyze the observed learning process and its interactions with the learning environment, and third, the emphasis on a strong social support system in an agentive learning environment led us to focus on socio-cultural networks in the observed engineering lab and their influence on individual lab members' learning and participation in research.

12.1.2 Learning, Self-regulation, and Feedback

The examples from the CLIC group show that learning within authentic research lab experiences empowers students, gives them agency over their learning processes and motivates them to build knowledge with others within the social and material learning space. Two aspects of such learning experiences are particularly important to us: self-regulation and feedback.

Self-regulation is essential to advance in any academic field, and cultivating self-regulated learners is one of the central goals in higher education (Hodges 2015). It is also a crucial element of the process of professionalization because in many workplaces, employees are expected to become responsible for their self-directed learning and training (Bell 2017; Cuyvers et al. 2020; Kittel et al. 2021; Smith and Curtis 2020).

Self-regulation is understood as an ability to monitor one's learning process and goals, to defer gratification, and to persevere in the face of difficulties. Self-regulation is linked to a set of beliefs about the quality of one's own knowledge and skills (self-efficacy) and to students' orientation toward achievement (mastery goals versus performance goals). Self-regulated learners are therefore aware of their knowledge, skills, beliefs, motivation, and cognitive processing and are capable of monitoring how well their cognitive engagement in learning matches their goals (Butler and Winne 1995; Van den Boom et al. 2007).

According to the evidence gathered by Zimmerman (2002; Zimmerman and Schunk 2001), self-regulatory and learning processes are intertwined in a cyclical way. At the onset of the learning process and during the *forethought* phase, learners' sense of self-efficacy and achievement orientation help define and set learning goals and the strategies to achieve them. As the learners get engaged in learning—the *performance* phase—the metacognitive aspects of learning, such as self-control and self-monitoring become evident. In the last phase of the self-regulatory process in learning, the *reflection* phase, the learners self-evaluate the results of their learning by reacting to them and by assigning causes to the learning outcomes that may or may not impact future learning strategies.

Completing a self-regulation cycle while learning depends on an individual learner's self-efficacy, achievement orientation, internal feedback, external feedback, and the interactions within and interconnectedness of a learning community (Hodges 2015). Butler and Winne (1995) point out that effective learners develop certain feedback routines while engaging with an academic task. For instance, learners may set a plan for working on a task, and such a plan, in turn, allows for generating criteria against which progress on a task is monitored. Students may modify their learning by readjusting present goals or setting new ones, and they may adapt existing skills or reevaluate and change learning strategies and decide upon more productive approaches. However, when a task at hand is too difficult or complex and its requirements exceed learners' current level of skills and knowledge, self-regulated learners actively seek feedback from external sources such as teachers' comments and peers' contributions in collaborative groups, and such external feedback often increases the effectiveness of learning.

As mentioned earlier, authentic research experiences in a research laboratory fall within the idea of pedagogies of engagement that seek to cultivate self-regulated and agentive learning. A research lab environment that supports agentive learning allows learners to interact freely with peers and mentors, and with artifacts, affording multiple opportunities for internal and external feedback. The emphasis in such an environment on empowering learners to stay motivated and to seek opportunities for their professional development affords the development of self-regulated learning.

12.2 Learning in Engineering Research Labs during the COVID-19 Pandemic

12.2.1 Cognitive Ethnography: Observing Learning as it Unfolds

Following CLIC's attempts to study learning processes in a biomedical engineering laboratory through cognitive ethnography, we implemented a similar methodological approach. Cognitive ethnography is a method aimed at investigating human cognition as expressed through cultural activity within real-world, "naturalistic" settings. It aims for understanding, explaining, or intervening in cognitive phenomena through detailed and refined analysis of ethnographic observation, digital photography, and video recordings. Hutchins argues for using cognitive ethnography to refine our understanding of how human cognition functions as "our folk and professional models of cognitive performance do not match what appears when cognition in the wild is examined carefully" (Hutchins 1995). Cognitive ethnography employs many of the same skills and practices as traditional ethnography, such as participant-observation, interviewing, and artifacts analysis. However, it combines them with the analytic techniques and theories derived from contemporary cognitive science. A central technique of cognitive ethnography consists of microanalysis of specific occurrences of events and practices to analyze in great detail the mechanisms and processes of interpersonal, naturally situated, and techno-socially distributed cognition (Alač and Hutchins 2004; Dubbels 2011; Hutchins and Palen 1997). To facilitate microanalysis, cognitive ethnography makes significant use of digital recording media and painstaking qualitative analysis of recordings (Williams 2006). Cognitive ethnography offers a powerful pathway towards understanding the functional specification of human cognition at both the individual and distributed scales; in other words, the functional role that cognition plays in relation to the environment, technology, society, culture, ongoing practices and activities, goals and problems, etc. Thus, cognitive ethnography is increasingly employed for studying cognitive activities that are environmentally and culturally situated (Lave 1988; Lave and Wenger 1991; Roth and Jornet 2013) and socially and technically distributed (Hutchins 1995 2014; Nersessian 2019; Nersessian and Newstetter 2013; Salomon 1993; Sutton 2006).

At the onset of our study on authentic research experience and learning in engineering labs, we used such traditional ethnographic tools as participant-observations, interviews, use of field notes, photos, and video-recordings. However, as the COVID-19 pandemic rendered our study impossible to continue, we decided to introduce an autoethnographic tool that our participant observer and our participants used to share accounts of their research and lab experiences.

Drawing from a cognitive ethnographic toolbox, we applied analytic frameworks from cognitive science, which enables fine-grained, micro-scale, and holistic analysis of cognitive activities that consist of discourses, nonverbal expressions, artifacts, and environment (Alač and Hutchins 2004; Grohman et al. 2020; Lee et al. 2015, 2017, 2020; Hutchins and Palen 1997; Williams, 2006).

12.2.1.1 What does a Participant Observer Do?

Our participant observer, one of the research team members, was embedded in the field site in January, 2020. The participant observer obtained access to a primary and a secondary laboratory and had a desk in the primary lab. She regularly visited the lab for hours at a time, observed the lab activities, asked questions, and took field notes. Occasionally, she conducted informal interviews. She also participated in a few lab activities. For example, she volunteered to be one of the subjects who tested a robot system, and she assisted to prepare another robot system for demonstration. When the COVID-19 pandemic hit, the participant observer had been collecting data in the field site for approximately two months.

At the beginning of the pandemic, all labs in the university were closed, and our field site was no exception. For a month, all research activities at the field site lab were halted. As all in-person data collection was prohibited, participant observation was no longer possible. After about a month, the lab members in the field site started holding lab meetings online, and we also began to observe these online meetings. The participant observer attended every lab meeting and conducted online interviews to collect data. The research team also obtained access to recordings for five of these meetings, and the lab members shared their research updates at the participant observer's request.

12.2.1.2 Participants as Observers: Collaborative Autoethnography

Participating in the lab meetings during COVID-19 lockdowns allowed us to collect some data, but without the benefits of participant observation. That is, we no longer could interact with the lab members in our field site by participating in their activities. On the other hand, the COVID-19 pandemic provided an unexpected opportunity to use an alternative ethnographic tool to interact with our participants, namely Collaborative Autoethnography (CAE; see Roy and Ueksa 2020 for a report on application in qualitative research). Collaborative Autoethnography is "(…) a qualitative research method that is simultaneously collaborative, autobiographical, and ethnographic (Chang et al. 2012)." It is based on self-reflection about an individual's experiences related to cultural, social, or physical circumstances. CAE also includes collaboration, allowing individuals to share and co-analyze their autoethnographic data to find meanings. By applying CAE in our study, we provided an opportunity for the participants and the participant observer to write and share their personal stories on the ways they coped with pandemic-related restrictions in relation to their research experiences. The CAE was also augmented by follow-up questions or interviews with the study participants.

12.2.2 Two Research Labs[1], Two Stories

University research laboratories are typically independently run by their lead faculty. Departments and schools offer minimal oversight, and faculty are largely able to shape the practices and culture as they see fit. Often these practices and cultures are not formally or intentionally imposed, but arise organically from the faculty member's leadership style, their personality, and those of the researchers, etc. Thus, no two labs will ever be the same.

12.2.2.1 MHR Lab

Our primary field site was a mechanical engineering research lab at a state university in Texas. The primary research projects in the MHR Lab revolved around human-robot sensory interactions during medical intervention. At the onset of our study, the lab was of moderate size and included seven regular members: the lab director (a faculty member at the university), a postdoctoral researcher who recently joined the lab, and five graduate students. However, during the course of our observations, the MHR Lab underwent a few changes, some related to restrictions and lockdowns due to COVID-19, and some due to personnel changes. Specifically, during the early months of the COVID-19 pandemic, the lab was closed and some lab activities shifted to the online environment. When the most severe restrictions were lifted and the MHR lab reopened, the lab members had to follow strict limited-capacity guidelines, meaning that the lab members had to take turns working remotely or in the lab (only one lab member was allowed to be in the laboratory space at the time). As far as the personnel changes are concerned, during the course of our observation, one graduate student obtained her Ph.D. and became a postdoctoral researcher. However, the major event that coincided with our observation was the lab director's and lab members' relocation to another state university in Texas.

12.2.2.2 AP Lab

The AP Lab is at the same state university in Texas as MHR Lab originally was. It is a material science and engineering lab that focuses on advanced polymer research. The decision to add AP Lab as a secondary field site was dictated by the COVID-19 pandemic restrictions on in-person data collection and by limited opportunities to observe the labs as they operated in an online environment. With the MHR Lab relocation to another university, the AP Lab became our primary field site. At the beginning of the data collection, the AP Lab included approximately 16 lab members including the lab director, two postdoctoral researchers, and graduated students (some of them interns at local companies). Out of this group, eleven lab members—a postdoctoral

[1] The names of the labs are not acronyms but pseudonyms created by the participant observer. The lab directors' and lab members' names are pseudonyms as well.

researcher and ten graduate students—have consented to participate in our study and have been regularly attending online lab meetings. Similarly to the MHR Lab, the work at the AP Lab has been restricted due to the COVID-19 pandemic guidelines; the lab members have had to alternate their on-site shifts with remote work.

When comparing the field notes on the two labs, the methods of engagement by the lab director became clear, which are partially the result of the differences between the labs' sizes. In the MHR lab, medium size research laboratory, the lab's director was involved in organizing and leading the lab activities. In the AP lab, a much larger lab, the role of the organizer of lab activities, including the meetings, was assumed by a postdoctoral researcher.

12.2.3 *The Impact of COVID-19 on the Research Labs*

To uncover the ways the COVID-19 pandemic affected the authentic research experiences in the two field sites, we analyzed the data included in the participant observer's field notes, video recordings of the online lab meetings, interviews with the lab members, and the CAE stories written by the lab members. Specifically, we focused our analyses on those observed events that included various types of interactions among the lab members, such as peer discussions, teaching and learning exchanges, and collective problem-solving sessions. Examining those interactions allowed us to identify various roles lab members had in the identified events, emerging learning goals, and types of learning activities. As far as the CAE stories are concerned, we examined the emerging patterns of experiences related to the pandemic and its effects on the lab members' professional and personal lives.

12.2.3.1 Lab Meetings

Before the COVID-19 pandemic, the MHR and AP labs had a history of regularly scheduled lab meetings, during which the lab members (mostly graduate students) presented their research updates and received feedback from their peers and faculty. But once the COVID-19 related university lockdown forced the labs to shift their activities to virtual spaces, the new role of the lab meetings emerged, and different meeting styles of the MHR and AP labs became apparent.

When observing the MHR Lab online meetings, we noticed the friendly and relaxed atmosphere among the lab members. They usually kept their video feed on during the meetings and engaged in discussing various topics related and unrelated to the MHR Lab's research projects. For example, the lab members talked about selecting new furniture for the lab, discussed a mentoring session on writing peer-reviewed research papers, collaborated on writing a review paper, or presented updates on their individual research projects. The AP Lab meetings had a different style and resembled formal meetings, during which the lab members followed an agenda that focused on lab and research-related issues. The meetings included announcements of and updates on the AP Lab activities, discussions of lab safety issues, and student presentations. The formal character of the meetings was coupled with the lab members' preference to keep the video feed off, except for the postdoctoral researcher and mentor who led the meetings, and the lab member appointed to present their research update (the MHR and AP Labs meeting styles are summarized in Table 12.1).

With the different meeting styles emerging from the participant-observations of the online MHR and AP Labs meetings, we solicited the lab members' opinions about the meetings. The follow-up interviews revealed that people in both labs thought of the online meetings as a good opportunity to engage with one another professionally and socially. This was particularly significant to MHR and AP labs members when the university partially lifted the restrictions and the labs could reopen with limited capacity. During the limited-capacity operation, the lab members had to alternate shifts and could not share the physical lab space with other people, so even though some of the research-related activities took place in the physical lab space, the lab meetings remained the only occasion to virtually meet all of the lab members at the same time. The lab members admitted that attending the online lab meetings gave them the sense of security and belonging so much that the online lab meetings continued even after the university and lab activities were restored to full capacity and in-person mode. It appears that during the COVID-19 crisis, the regular online meetings were not only a venue for the research-related exchanges, but a new role as an anchor to hold the lab members together as a group also emerged.

Table 12.1 Comparison of online lab meetings in two field sites

Meeting characteristics	MHR Lab	AP Lab
Frequency and duration	Weekly, 1 h	Weekly, 1–1.5 h
Organizer	Lab Director (faculty)	Post-Doc Mentor
Participants	2 post-docs, 4 graduate students	10 graduate students
Video conference	Lab Director: on video All participants on video	Post-Doc Mentor: on video Participants: off video Presenter: on video
Topic to discuss	Various topics Decide what to discuss next at the end of the meeting	Fixed topics: Lab updates; Lab safety issues; Student research update presentation
Procedure	Varied according to the topic	(1) Lab announcement and updates (2) Lab safety discussion (3) Student Presentation (1 or 2)

12.2.3.2 Lab Members' Stories

Four members from the MHR Lab and seven members from the AP Lab shared written CAE stories. To analyze these stories, we developed an analytic framework based on Levine et al.'s (2021) study report that examined the impact of the COVID-19 pandemic on early career scholars and doctoral students in the education area. Adjusted to the unique background and circumstances of graduate students in engineering, our analytic framework included the following themes that emerged from the key terms the labs members used in their stories: background, impacts on research, continuing research work, learning, connections/community/communication, and work-life balance. Each story was carefully analyzed based on these themes, and the general profile for each theme was extracted. The following examples are excerpts from the CAE stories written by the MHR and AP lab members. The findings are summarized in Table 12.2.

The analysis of the emerging themes in the written CAE stories suggests that the members in our participating labs were trying to continue research activities as best they could during the COVID-19 pandemic and the related lockdowns. The *Background* theme indicates that at the onset of the university lockdown, the MHR and AP Labs members were mostly planning or preparing for a new research project, graduation, or internship. As the COVID-19 pandemic effectively stopped them in their tracks, the lab members experienced and shared negative impacts of the pandemic on their research activities that were now *delayed*, *slowed*, or had *suffered in efficiency* (see Table 12.2). Despite the difficulties the lab members encountered, they tried to continue their research and learning or shift to alternative activities and remain productive or beneficial in a wider sense, such as joining an expert group to help health workers.

The themes related to continuing research work allowed us to notice an interesting difference between the MHR and AP labs, perhaps based on their areas of expertise (see Table 12.2). The MHR Lab focuses on human-robot interaction and their research protocols involve human subjects. Thus, when in-person data collection was suspended, the MHR lab members, either shifted their research activities to nonhuman subject studies such as virtual simulation or used their time to revisit and reevaluate previous data analyses. While shifting the focus of research activities away from in-person data collection, the MHR lab members continued to work remotely. In contrast, the area of expertise among the AP Lab members is in material science and does not require interactions with human subjects. Therefore, many AP Lab members tried to go back to the lab by obtaining permission to do so or by adhering to the limited-capacity guidelines (e.g., social distancing, taking turns to work in the lab alone). For example, one lab member took night shifts to work in the lab and, consequently, they did not see any of the fellow lab members for months! Apart from seeking ways to continue work in the laboratory space, quite a few AP Lab members joined a local expert group to develop innovative personal protective equipment (specifically masks) to help frontline healthcare workers. This way, not only did they continue their engagement in authentic research experiences, but they also made a contribution with their expertise in material science.

Table 12.2 Key themes and key words in the lab members' CAE narratives with examples

Key themes	Key words		Examples
MHR/AP Labs	MHR Lab	AP Lab	MHR/AP Labs
Background	Start, prepare	Plan, graduate, prepare	*"Things were going well and I was just about to get started on human studies for the sleeve* [surgical tactile device—MG]. *when the Covid-19 situation started"* (Rob, post-doc, MHR Lab)
Impacts on research	Difficult to manage, suffered in efficiency, no human studies	Affected my (plan, graduation, work pace), delayed, slowed	*"(...) by the end of May, I started to notice a little slow down in my activities. This* [new] *routine began to affect my work pace"* (Oliver, Ph.D. student, AP Lab)
Continuing research work	Shifting, revisiting, reevaluating	Joined (expertise group to help, a new group to continue), get back to lab (with special permission, with limited capacity)	*"I decided to change my thesis topic again, to try making better facemasks (...) to at least try to do something"* (Brad, AP Lab)
Learning	Learning, practicing	Learning (a new way, software, etc.), reading and reviewing papers	*"I had reached a point in my research that allowed me to focus on data analysis and paper writing, so I wasn't worried about continuing my research"* (Rob, post-doc, MHR Lab)

(continued)

Table 12.2 (continued)

Key themes	Key words		Examples
MHR/AP Labs	MHR Lab	AP Lab	MHR/AP Labs
Connection-community-communication	Same communication in professional life	Keep us accountable, keep in touch	*"I miss to have the opportunity to practice my presentations in person but I believe that having the opportunity to keep in touch with the lab members has been good for all the group members, especially for those who live alone"* (Caro, Ph.D. candidate, AP Lab)
Work-life balance	Challenges in home environment, use the commuting time for something else	Not working very well (with two small kids), not that bad (used time for personal growth)	*"(...) working from home has its challenges. I live in a one-bedroom apartment, which I share with my wife, two cats and a dog, and I now* [wish] *that we had gone with a two-bedroom unit, so that I could be working from an office of sorts, instead of a table in my living room".* (Rob, post-doc, MHR Lab)

The analysis of the theme related to personal circumstances and work-life balance revealed yet another difference among the lab members. For example, two members of the AP Lab expressed different feelings about balancing professional and personal life during the pandemic. A member who was a parent with two small kids mentioned "not working well" and feeling "exhausted," while another member who lived alone mentioned, that balancing professional and personal life was "not that bad" because he could use the time for personal growth (see Table 12.2).

12.2.4 Emergent Learning Processes in the Research Labs

12.2.4.1 Self-regulated Learning as the Default Learning Strategy

Based on our observations and analyses of in-person and online interactions between the lab members in the MHR and AP Labs, we propose that the learning in the engineering research labs is agentive and self-regulated (Hodges 2015; Nersessian and Newstetter 2013; Newstetter et al. 2002, 2004; Zimmerman 1990; Zimmerman and Schunk 2001), and as such indicates that learners take control of their own learning process and remain motivated and empowered to conceptualize, design, and execute learning tasks related to their research activities. The members of the MHR and AP Labs each had their own research project to conduct. While they could seek guidance or help from their advising faculty or more experienced peers to choose their topic, plan the research procedure, or to solve any emerging problems, the decisions were their own to make.

As illustrated in the cultural model of MHR and AP lab members' self-regulated learning processes (Figure 12.1), the lab members chose what they wanted to learn and made plans about how to proceed. They controlled the procedure and revised or altered it as needed. These self-regulated learning processes were closely related to the individual lab members' research projects and were still clearly observable when the lab members were meeting online during the COVID-19 pandemic. As indicated in the analysis of the CAE stories, to adjust to the pandemic-related restrictions, the lab members actively sought alternative ways to continue their research. Some switched their research topic to a topic that they could do at home. If possible, they took all the devices and resources home and continued to work, or where it was not possible to continue their initial research projects, they participated in alternative projects, such as using their expertise to combat the pandemic.

12.2.5 Feedback and Scaffolding as Organic Instructional Strategy

Self-regulated learners actively seek feedback from external sources (e.g., teachers' and peers' comments) and, in turn, the external feedback enhances self-regulated learning (Butler and Winne 1995). During the observation of the MHR Lab, we noticed that the lab director—who was also an advising faculty to graduate students—provided feedback that helped to guide self-regulated learning in the lab. Namely, the lab director differentiated her feedback according to the current level of skills and abilities of the student who presented their work during the lab meetings. When the presenter was experienced and familiar with the topic, the lab director asked questions or made comments that tend to trigger more discussion about the content of the presentation such as "Have you tried ...?" and "One idea I have about this is ...". Thus, the exchange between the lab director and the more experienced student resembled

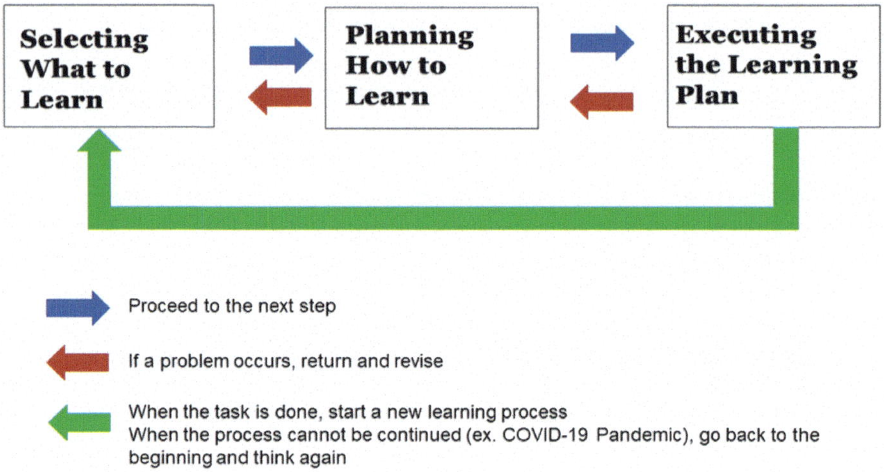

Fig. 12.1. The cultural model of self-regulated learning process that occurs in the engineering lab

discussions between professionals at an academic conference. However, when the presenter was less experienced, or the topic was unfamiliar to the presenter, the lab director often provided additional explanations—"Let me explain…"—or brought up related knowledge—"Do you remember ...?" In contrast to the quasi-professional exchanges described above, the back and forth here resembled interactive teaching approaches in the classroom.

The way the lab director used feedback during the student presentations (i.e., considering students' skills, knowledge, and experience) reminded us of the pedagogical strategy known as "scaffolding" (Grohman et al., in preparation). In general, scaffolding is a strategy in which an expert provides the necessary support for a learner to accomplish a specific task, and this support is differentiated according to the learner's ability and situation. As the learner obtains more independence, the expert's support is gradually diminished (Malero et al. 2012; Sharma and Hannafin 2007; Wood et al. 1976).

In most educational settings, scaffolding is used as an intentional or overt instructional strategy to guide the learning process (Malero et al. 2012). Interestingly, the lab director was not aware that she was using scaffolding as an instructional strategy, as shown in the following excerpt from her email correspondence:

> I'm not aware that I'm doing it, but I do try to make sure that if a student brings something up that only the two of us are familiar with, I try to give that context to the group. Also, as students get more experienced, I try to encourage them to act more independently since that is my long-term goal for their education (Dr. Mac, the director of the MHR Lab).

As can be seen in the excerpt, scaffolding used in the MHR Lab occurs without formal planning or instructional design. We call this type of instructional strategy an *organically-occurring scaffolding*. We discuss more details about scaffolding in the engineering lab in a separate publication.

12.3 Implications for Professionalization and Learning in Research Labs during Crises

What are the larger implications for professionalization and education during a crisis like the COVID-19 pandemic? First, it is important and hopefully not too obvious to say: students do not stop learning in a pandemic lockdown, despite the incredible disruptions to routine. Under those conditions, labs had to develop new mechanisms for supporting learning mechanisms such as self-regulation and feedback. In particular, we noted that self-regulation became a more significant factor than it might be under normal conditions. It is not a stretch to hypothesize that prior preparation for self-regulatory learning might be an advantage in many disruptive scenarios, giving students with such skills greater resilience.

Regular laboratory meetings became a forum for both professional and social support. Under pre-pandemic circumstances, students received social support inside and outside the laboratory through their everyday interactions with colleagues and friends. Such casual interactions were limited for all of us who were engaged in some form of sheltering or isolation during the pandemic. In the laboratory in particular, casual interaction almost entirely ceased. The two laboratories we observed reacted very differently to these changes. Lab meetings under pre-pandemic conditions typically had a more or less formal structure that helped them to efficiently serve their function of exchanging information, planning, and learning. The AP lab retained the basic structure and tenor of those meetings during the pandemic.

The MHR lab, on the other hand, adopted a less formal, more "friendly" approach, and in so doing became a major forum of social support for the members. This seems to have led to a more fluid transition for the MHR lab to shifting work focus and addressing the problems at hand. There was greater engagement of the members in the meetings, including most members having their cameras on, and explicit provision of social support. Through these meetings, the MHR lab provided a social support network for its members that eased the difficulties created by the pandemic response.

The COVID-19 pandemic gave us a rare opportunity to observe how student learning, professionalization, and research experience are constrained by their individual circumstances. Learning occurs in contexts that go beyond the university classroom and the laboratory. It is difficult to overstate the importance of this insight; educators often wrongly understand student learning as entirely a function of their performance in classroom and laboratory activities. But the students' educational and professional goals and outcomes are in fact intertwined with their private lives and the resources they have at hand. In this case, social support from lab mates, work from home arrangements, and availability of material resources at home, all had important impacts on the students' learning and professionalization.

A key insight of distributed cognition theory (Hutchins 1995) is that social relations, social organization, material artifacts, and social and communicative acts can be not only *about* cognitive matters; they can *be* cognition. Nevertheless, for most of us, under most conditions, the *social* meaning of our relationships, organizational structures, words, and actions are most salient. For Hutchins, this is merely a reflection of our evolved capacity as social creatures; as he says, "If it is true that human minds evolved to process social relations, then packaging a task in a social organization may facilitate understanding it" (1995, p. 263). The cognitive activities of learning and knowledge-creation are packaged in the social organization of the laboratory; those activities will be disrupted if the laboratory no longer can support the social relationships and meanings it previously provided. "Social moves have [cognitive] as well as social consequences. [Cognitive] moves have social as well as [cognitive] consequences" (ibid.).[2]

12.4 Note to Our Future Colleagues

Greetings from your colleagues in the academic year 2021-22. We are currently in, what we hope is, the tail end of the COVID-19 pandemic. Hopefully, it recedes in the next year or so, and life—including university education and research—becomes similar to what it was prior to the pandemic. However, we recognize that this could be much more drawn out, and that COVID-19 could linger for years, with potential future outbreaks. Over the next 30 years, it seems a near certainty that there will be other severely disruptive events. For example, we already see extreme weather events forcing closures and evacuations for a week or more at a time. Climate change will likely exacerbate this and lead to other problems. We believe that our study offers invaluable insights into how to guide learning and professionalization in research laboratories during disruptive and adverse events.

In our research during the COVID-19 pandemic, we recognized the resilience and drive of researchers at all levels. From senior faculty to undergraduates, research finds its way to thrive. The two labs we observed pursued different strategies to continue their work, as befit their research focus, the experience of their members, and size. What the students across AP and MHR labs had in common was a notable level of self-regulation of their own learning, including the ability to adapt to changing circumstances in their professional and personal lives. This self-regulated learning was accompanied by feedback from senior lab members and directors adjusted to the learners' research experiences and their personal contexts. This plasticity in self-regulation and feedback may be a necessary component of robust research labs during uncertain times.

Another common characteristic in the two labs was a need to maintain a social "safety net" through building and cultivating collegial relationships with the lab members. The social cohesion was facilitated by videoconferencing tools, which are now quickly becoming a "new normal" method of formal research and education, even as universities and other workplaces ease restrictions on in-person attendance.

---

[2] In the original text, Hutchins says "Computational" instead of "cognitive"—in the book he relies on the then-standard computational-representational understanding of "cognition" and moves easily between the two, in a way that we might be less comfortable with today.

We predict that this trend will continue over the next 30 years. Many companies are already experimenting with permanently remote workers.

The ability to conduct research remotely, at the level of a Master's thesis or Ph.D. dissertation, has not been well-explored. It now seems feasible that graduate research degrees could be awarded through purely remote access, or through a mixture of remote and in-person work in the lab. This can allow work to continue during future disruptions and open the way to graduate STEM education to students who would previously not be able to pursue an advanced degree due to difficulties in the residential requirements of graduate education. One issue that is clear to us, is the need to foster the spirit of collegiality within the lab. Lab members will continue to need social outlets and interactions, even if just during weekly meetings, and research laboratories will require leadership invested in encouraging and maintaining a productive and collegial atmosphere among their students, staff, and colleagues. The material and social environment of classrooms and laboratories cannot be ignored as we design future learning, professionalization, and work environments and processes.

Acknowledgements The study discussed here was supported by NSF Award 1902072, *The formation of engineers in the research lab: A cognitive ethnography study.*, September, 2019–August 31, 2022.

References

Alač M, Hutchins E (2004) I see what you are saying: Action as cognition in fMRI brain mapping practice. J Cogn Cult 4:629–661

Bell BS (2017) Strategies for supporting self-regulation during self-directed learning in the workplace. In Ellingson JE, Noe RA (eds) Autonomous learning in the workplace. Routledge/Taylor & Francis Group, pp 117–134. https://doi.org/10.4324/9781315674131-7

Butler DL, Winne PH (1995) Feedback and self-regulated learning: a theoretical synthesis. Rev Educ Res 65:245–281

Chang H, Ngunjiri F, Hernandez KC (2012) Collaborative autoethnography. Retrieved from http://ebookcentral.proquest.com. Created from UTD on 2020-03-30 15:20:10

College Crisis Initiative (n.d). Retrieved from https://collegecrisis.shinyapps.io/dashboard/

Cuyvers K, Van den Bossche P, Donche V (2020) Self-regulation of professional learning in the workplace: a state of the art and future perspectives. Vocat Learn 13:281–312

Dewey J (1933/1998) How we think: a restatement of the relation of reflective thinking to the educative process. Houghton Mifflin, Boston, MA

Dewey J (1938) Experience and education. Kappa Delta Pi

Dubbels B (2011) Cognitive ethnography: A methodology for measure and analysis of learning for game studies. Int J Gaming Comput-Med Simulat 3:68–78

Edgerton R (1997) Education white paper. Unpublished report for the Pew Charitable Trust

Frantz KJ, DeHaan RL, Demetrikopoulos MK, Carruth LL (2006) Routes to research for novice undergraduate neuroscientists. CBE Life Sci Educ 5:175–187

Giere RN, Moffatt B (2003) Distributed cognition: where the cognitive and the social merge. Soc Stud Sci 33:301–310. https://doi.org/10.1177/03063127030332017

Grohman MG, Gans N, Majewicz Fey A, Lee EA, Brown MJ (in preparation). Scaffolding in the engineering lab: a strategy that enriches learning in research experience

Grohman MG, Gans N, Lee EA, Tacca M, Brown MJ (2020) Engineering ethics as an expert-guided and socially-situated activity. Adv Eng Educ 8(3):1–16. https://advances.asee.org/engineering-ethics-as-an-expert-guided-and-socially-situated-activity/
Hodges LC (2015) Teaching undergraduate science. Stylus Publishing, Sterling, VA
Hutchins E (2014) The cultural ecosystem of human cognition. Philos Psychol 27:34–49
Hutchins E (1995) Cognition in the wild. MIT press, Cambridge, MA
Hutchins E, Palen L (1997) Constructing meaning from space, gesture, and speech. In: Resneck LB, Saljo R, Pontecorvo C, Burge B (eds) Discourse, tools and reasoning: essays in situated cognition. Springer, Berlin, pp 23–39
Kittel AFD, Kunz RAC, Seufert T (2021) Self-regulation in informal workplace learning: Influence of organizational learning culture and job characteristics. Front Psychol 12:643748. https://doi.org/10.3389/fpsyg.2021.643748
Knorr Cetina K (1995) Laboratory studies: the cultural approach to the study of science. In: Jasanoff S, Markle GE, Peterson JC, Pinch T (eds) Handbook of science and technology studies. SAGE Publications Inc, Thousand Oaks, CA, pp 140–167
Knorr Cetina K (1999) Epistemic cultures: how the sciences make knowledge. Harvard University Press
Lang JM (2016) Small teaching. Everyday lessons from the science of learning. Jossey-Bass, San Francisco, CA (A Wiley Brand)
Latour B, Woolgar S (1979/1986) Laboratory life. The construction of scientific facts. Princeton University Press
Lave J (1988) Cognition in practice. Cambridge University Press, Cambridge, UK
Lave J, Wenger E (1991) Situated learning: legitimate peripheral participation. Cambridge University Press, Cambridge, UK
Lee EA, Gans N, Grohman M, Tacca M, Brown MJ (2020) Guiding engineering student teams' ethics discussions with peer advising. Sci Eng Ethics. https://doi.org/10.1007/s11948-020-00212-6
Lee EA, Grohman MG, Gans N, Tacca M, Brown MJ (2015) Exploring implicit understanding of engineering ethics in student teams. In: Proceedings of ASEE annual conference & exposition. Seattle, WA. https://peer.asee.org/exploring-implicit-understanding-of-engineering-ethics-in-student-teams
Lee EA, Grohman MG, Gans N, Tacca M, Brown MJ (2017) The roles of implicit understanding of engineering ethics in student teams' discussion. Sci Eng Ethics 23:1755–1774. https://link.springer.com/article/10.1007/s11948-016-9856-0
Levine FJ, Nasir NS, Rios-Aguilar C, Gildersleeve RE, Rosich KJ, Bang M, Bell NE, Holsapple MA (2021) Voices from the field: The impact of COVID-19 on early career scholars and doctoral students [Focus group study report]. American Educational Research Association, Spencer Foundation. https://doi.org/10.3102/aera20211
Linn MC, Palmer E, Baranger A, Gerard E, Stone E (2015) Undergraduate research experiences: impacts and opportunities. Science 347(6222). https://doi.org/10.1126/science.1261757
Madan ChR, Teitge BD (2013) The benefits of undergraduate research: the student's perspective. Mentor Acad Advisor J. https://dus.psu.edu/mentor/2013/05/undergraduate-research-students-perspective/
Malero J, Hernandez-Leo D, Blat J (2012) A review of constructivist learning methods with supporting tooling in ICT higher education: defining different types of scaffolding. J Univ Comput Sci 18:2334–2360. https://doi.org/10.3217/jucs-018-16-2334
Murdoch-Eaton D, Drewery S, Elton S, Emmerson C, Marshall M, Smith JA, Stark P, Whittle S (2010) What do medical students understand by research and research skills? Identifying research opportunities within undergraduate projects. Med Teach 32:e152–e160
National Research Council (2011) A framework for K-12 science education: practices, crosscutting concepts, and core ideas. The National Academies Press, Washington, DC
Nersessian N (2006) The cognitive-cultural system of the research laboratory. Organ Stud 27:125–145

Nersessian N (2009) How do engineering scientists think? Model-based simulation in biomedical engineering research laboratories. Top Cogn Sci 1:730–757
Nersessian N (2019) Interdisciplinarities in action: cognitive ethnography of bioengineering science research laboratories. Perspect Sci 27:553–581
Nersessian N, Newletter W (2013) Interdisciplinarity in engineering research and learning. In: Jori A, Olds B (eds) Cambridge handbook of engineering education research
Newsletter W, Kurz-Milcke E, Nersessian N (2004) Agentive learning in engineering research labs. In: Proceedings of 34th ASEE/IEEE frontiers in education conference, 20–23 Oct 2004, Savannah, GA
Newsletter W, Nersessian N, Kurz-Milcke E, Malone K (2002) Laboratory learning, classroom learning : looking for convergence/divergence in biomedical engineering. In: Proceedings of the international conference on learning sciences, Hillsdale, NJ. Lawrence Erlbaum, pp 315–321
NGSS Lead States (2013). Next generation science standards: for states, by states. http://www.nextgenscience.org/
Roth W-M, Jornet A (2013) Situated cognition. Wires Cogn Sci 2013(4):463–478. https://doi.org/10.1002/wcs.1242
Roy R, Ueksa S (2020) Collaborative autoethnography: "self-reflection" as a timely alternative research approach during the global pandemic. Qual Res J 20:383–392
Salomon G (1993) No distribution without individuals' cognition: a dynamic interactional view. In Salomon G (ed) Distributed cognitions: psychological and educational considerations (pp 111-4–128-5). Cambridge University Press, Cambridge, UK
Sharma P, Hannafin MJ (2007) Scaffolding in technology enhanced learning environments. Interact Learn Environ 15:27–46
Slovacek S, Whittinghill J, Flenoury L, Wiseman D (2012) Promoting minority success in the sciences: the minority opportunities in research programs at CSULA. J Res Sci Teach 49:199–217. https://doi.org/10.1002/tea.20451
Smith C, Curtis LJ (2020) Scaffolding learners' self-regulation in workplace learning: design of a pedagogical pattern for placements. J Vocat Educ Train 72:396–414
Sutton J (2006) Distributed cognition: domains and dimensions. Pragmat Cogn 14:235–247
Thompson RJ Jr (2014) Beyond reason and tolerance: the purpose of higher education. Oxford University Press
Van den Boom G, Paas F, Van Merriënboer JJG (2007) Effects of elicited reflections combined with tutor or peer feedback on self-regulated learning and learning outcomes. Learn Instr 17:532–548. https://doi.org/10.1016/j.learninstruc.2007.09.003
Williams RF (2006) Using cognitive ethnography to study instruction. In: Proceedings of the 7th international conference of the learning science. Lawrence Erlbaum Associates, Mahwah, NJ
Wood D, Bruner JS, Ross G (1976) The role of tutoring in problem solving. J Child Psychol Psychiatry 17:89–100
Zimmerman BJ (2002) Becoming a self-regulated learner: an overview. Theory into Practice 41:64–70. https://doi.org/10.1207/s15430421tip4102_2
Zimmerman BJ (1990) Self-regulated learning and academic achievement: an overview. Educ Psychol 25(1):3–17. https://doi.org/10.1207/s15326985ep2501_2
Zimmerman BJ, Schunk DH (eds) (2001) Self-regulated learning and academic achievement: theoretical perspectives, 2nd edn. Routledge. https://doi.org/10.4324/9781410601032

Chapter 13
Informal Learning of Science at Science Centers and Museums: Perspectives, Influences, and Issues

Elisa Hui Shi Chng and Timothy Ter Ming Tan

13.1 Introduction

The transition from the Industrial to the Information age has profoundly impacted society, necessitating new skills for citizens to adapt to an ever-changing future (Lai and Viering 2012). In more recent decades, these skills are often referred to as twenty-first century competencies. There is also a need for workers with broad-based science, technology, engineering, and mathematics (STEM) skills as they underpin research and innovation across multiple aspects of society that will enhance a nation's competitiveness (Martin et al. 2016). However, in the last two decades, there is a general trend of students in some industrialized nations no longer choosing to major in STEM subjects. This poses a concern as enrolment and performance in science at school has always been a vital foundation for a nation's growth (Martin et al. 2016; Suter 2014). It is thus generally acknowledged that finding ways to reverse the decline in STEM enrolments and student interest in science is important.

Informal science learning by its nature, can help build interest in science and foster science learning, and therefore support the learning of formal science curricula in schools. It could perhaps also help to address student interest and motivation in entering STEM-related fields of study and subsequently, professions. Science museums and similarly organized science places-of-interest are familiar venues to most people as places to experience and perhaps be entertained by science in a context that is generally unlike formal school science. It is therefore worth examining the literature in this area to understand what is known and how informal learning in these contexts can be further developed.

E. H. S. Chng
STEM Inc., Science Centre Singapore, Singapore, Singapore

T. T. M. Tan (✉)
Natural Sciences and Science Education Academic Group, National Institute of Education, Nanyang Technological University, Singapore, Singapore
e-mail: timothy.tan@nie.edu.sg

Y. S. Ong et al. (eds.), *A Diversity of Pathways through Science Education*,
https://doi.org/10.1007/978-981-97-2607-3_13

Science centers and museums (SCMs) have always been seen as repositories, charged with collecting, preserving, and curating items for academic study as their primary function. The exhibition of their collections were perhaps a natural extension, to display their wonders to a curious public, or even a necessity, to bring in needed funds. In time, SCMs are now identified as spaces with opportunities to enhance visitors' knowledge and motivation in science, and to highlight the connection between science and everyday life. Given the unequal distribution of science participation, opportunity, and aspirations across different strata of society and socio-economic status (Archer et al. 2014), publicly accessible sites (such as SCMs) that may promote science knowledge and motivation may be especially important in broadening access and equity in science participation and pathways.

In contrast to the formal learning of science in the structured curriculum of school settings, informal learning in science inherently involves flexibility. But this does not mean informal science learning is random or capricious. Such informal learning is typically self-directed, intentional, and interest-driven (Eshach 2007; Laurillard 2009). This has been helped by SCMs evolving over the years to include interactive elements in their exhibits (Crowley et al. 2014), engaging visitors through participatory and activity-based learning. The value of SCMs as places of learning also comes from their affordances for and fostering of curiosity and interest in science, development of inquiry skills, and opportunities for the general public to talk about science. In addition, schools can cooperate and work in conjunction with SCMs to provide opportunities for students to learn in informal learning spaces and exposing students to different perspectives and experiences.

The area of research concerning learning within informal settings has been on the rise since the beginning of the century, particularly with respect to visitors' learning and interactions when attending exhibitions, staff members' ideas about exhibit design, and other aspects such as the interaction between formal institutions and informal learning environments. This chapter aimed to summarize and highlight trends in the research that has been conducted in the two decades prior on informal learning of science in SCMs. Of particular interest were:

- Experiences and learning by different groups of visitors
- Interactions and engagement between visitors, exhibits and museum staff
- Factors influencing informal learning of complex science concepts
- Issues in the design, development, and presentation of exhibits

13.2 Methods

The literature search for this study was conducted using the preferred reporting items for systematic review and meta-analysis (PRISMA) protocol (Moher et al. 2015).

Step 1 Literature review. Initial search conducted using ERIC database for papers published between 2000 and 2020, both years inclusive. A Boolean keyword search using 'science museum', 'science center', or 'science centre' and 'science education' (see Fig. 13.1).

Step 2 The scope of this review included only empirical papers that had adopted rigorous research design, had been peer-reviewed, and published in high-quality, high-impact journals. Only papers from the two leading journals on museum education and four science education research journals listed in Table 13.1 were selected for inclusion.

Step 3 A total of 1,198 papers were retrieved using the Boolean search terms. Studies (n = 667) carried out before Year 2000 were considered outdated for this review. Another 418 papers were excluded because they were irrelevant (i.e., not in the context of SCMs). The remaining 113 papers were deemed to be relevant to this review. Non-empirical papers such as theoretical papers, positional papers, editorials, book reviews, and commentaries were excluded.

The selected 113 empirical studies are listed in the Bibliography section. Most used qualitative research methods (n = 56, 49.6%) and most of these classified by the authors as descriptive. Often, case-study designs (e.g., microanalysis or multiple case studies) or action research designs were used. Qualitative data collection included pre/post interviews, field notes and participatory observations of activities and interactions. Quantitative papers (n = 12, 10.6%) selected had to fulfil the criteria of employing large sample size and obtaining data through statistical analysis. Mixed methods papers (n = 45) accounted for the remaining 39.8% of the total.

The majority of the studies were conducted in North America and Europe (n = 85, 75.2%). This might be expected given the preponderance of science education

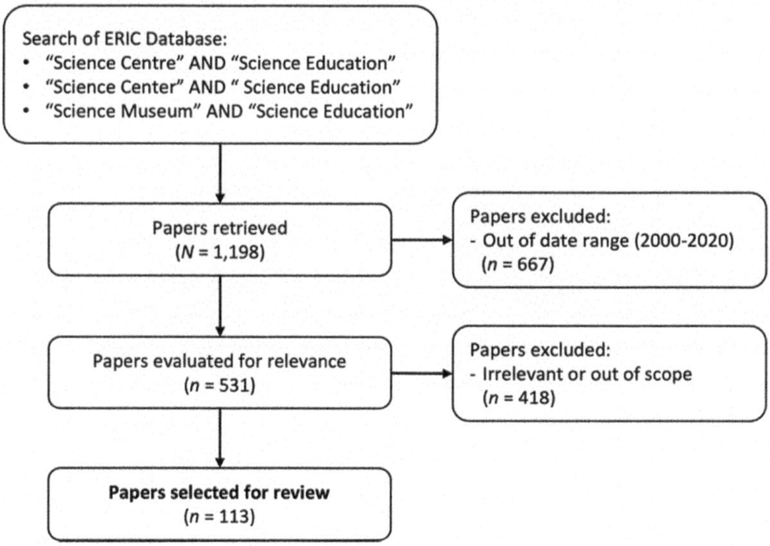

Fig. 13.1 PRISMA flowchart

Table 13.1 Source of papers by journal

Journal	Papers (*n*)	Percentage (%)
Cultural Studies of Science education	9	8.0
International Journal of Science Education & International Journal of Science Education (B)	44	38.9
Journal of research in science teaching	11	9.7
Journal of Museum Education	13	11.5
Research in Science Education	7	6.2
Science Education	29	25.7

research in general from industrialized nations. However, this highlights the under-representation of papers ($n = 6$, 5.3%) from highly populated regions such as Asia, and hence the opportunities for research in other contexts and perspectives. Almost two-thirds of the papers were from just two journals (see Table 13.1).

The selected papers were examined to determine their research aims and central findings, as well as any discursive content of interest. These key points were then grouped into emergent thematic similarities with the aim of identifying the key research foci in the field of informal science learning, in the context of SCMs. Categorization of the key points was done iteratively through discussion between the authors to define a coherent hierarchy which also reflected the natural interactional relationships between the themes. The key themes (boxes) and their relationships (arrows) are summarized in Fig. 13.2. The rest of this chapter has been organized with the identified common themes as the headings and sub-headings under which exemplars of the research or research findings will be discussed.

13.3 Analysis and Discussion

The underlying and often unstated assumption in all papers is that the main intended outcome of a visitor entering a science center or museum is for them to take away some learning of, or at least develop their awareness of or interest in, science—in terms of scientific knowledge, the nature of the scientific endeavor, and/or socio-scientific issues. However, it is perhaps not surprising that investigating how visitors learn in museums is a challenge due to a multitude of interacting factors (e.g., motivation and beliefs, design of exhibition, and social interaction) (Falk and Dierking 2000).

Most studies of SCMs relate to understanding visitors and their experiences, the design and effectiveness of exhibits/programs, or what can be collectively considered the work of museum staff as well as the role and operation of SCMs. Visitor-related studies focus on the visitors' behavior, actions, and dialogues, which are usually related to the specific context of the exhibition. Program and exhibit-related studies

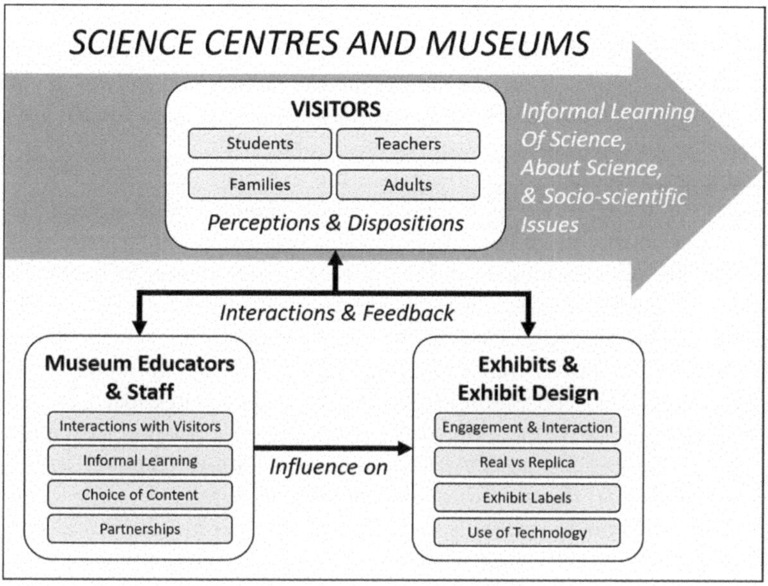

Fig. 13.2 Key themes and relationships identified in this review

often seek to explore and evaluate the effectiveness of exhibit content and design in relation to visitors' learning. The last major group of studies examined a range of topics relating to the work of museum staff, including papers on the theoretical perspectives that underpin informal science learning in SCMs and factors that influence museum educators' exhibition design decisions.

13.3.1 Visitors

In this section, the perceptions from each visitor group are discussed separately as each group provides researchers with different insights. Visitors are categorized into four groups: students, teachers, families, and adult visitors. Visitors to SCMs are mostly students and students of all ages are grouped together, from preschool to university students. The grouping of teachers includes in-service and pre-service teachers and museum educators. Family visitors consist of at least one parent and child, while adult visitors constitute all other general visitors visiting the SCMs.

13.3.1.1 Students

Studies on students' perception and learning in a science center can be separated into two groups, where one group focuses on the general and cognitive learning processes in students, while the other group of studies focuses on the social and emotional aspects of student learning.

Studies like Shaby et al. (2019a) sought to measure how much students could recall from exhibits and their museum experience. They noted four factors that significantly affect student learning: freedom to explore the exhibits without a structured program; having sufficient technical skills to operate the exhibits; facilitation by museum educators (e.g., guided tours, explanations), and social interactions between students. The correlation between freedom of choice and better conceptual learning is also known in students with learning disabilities (Brooke and Solomon 2001). Scaffolding of learning in science museums was also investigated by Yoon et al. (2013) and Krange et al. (2020). One key observation from Yoon et al. (2013) is that the amount of scaffolding provided was inversely related to the amount of exploratory behaviors.

Walker et al. (2013) and Salmi et al. (2017b) looked into aspects of 'edutainment'. Walker et al. (2013) examined the use of science theatre presentations to address HIV-AIDS issues among African students to motivate change in behavior, while Salmi et al. (2017b) used dinosaur exhibits to investigate students' interest and motivation towards science during learning. Salmi et al. (2017b) also noticed an effect of autonomy on learning attitudes, with greater autonomy in exploring the exhibits leading to better learning outcomes.

As for studies investigating the social and emotional aspects of student learning, Şentürk and Özdemir (2014) examined students through the use of attitude scales before and after a science museum visit. Results showed that students' attitudes towards science improved immediately after the visit but declined one week later. Further analysis of their results showed a gender effect on attitudes, with females having a greater attitudinal changes (larger improvement in attitude, but also having a larger decline in attitude) as compared to males. Gender effects in students' learning (Şentürk and Özdemir 2014; Salmi et al. 2017b) have also been confirmed by Silfver (2019). In all, they conclude that science attitudes are highly influenced by the way science concepts are presented (see later sections for discussions on *sociocultural perspectives*).

From the social aspect of informal science learning, Burke (2020) investigated dispositions of students from low-income families towards informal science education and suggested that science educators may be unintentionally limiting the children's learning opportunities in science (e.g., by hiding the word "science" from children), and recommends changing the perceptions of informal science educators, so that they can engage the children better (see discussion under the section for the *Families* group of visitors, below).

Measuring students' learning and perception in SCMs is a complex task as there are many elements that may influence the outcome of their learning. However, Suter (2014) has shown that continued visits to SCMs allowed students to develop *science capital*, as observed from higher achievements in science and mathematics.

13.3.1.2 Teachers

Research on this group generally focuses on the roles and perceptions of teachers who visited SCMs with their classes. These papers aimed to understand the roles teachers play in class visits to SCMs. There are several reasons for taking school children to museums: science enrichment, increasing student exposure to science, social experience and entertainment, and having concrete experiences with abstract and complex science phenomena (Griffin 2004).

Teachers generally have a good impression of SCMs, as they perceive SCMs as places that allow students to gain new knowledge by allowing them to have hands-on experiences with science concepts (Tal et al. 2005; Karnezou et al. 2013; Faria and Chagas 2013). Luehmann and Markowitz (2007) reported that as teachers became more familiar with SCMs, they gradually shifted their focus from students' grades to students' motivation and science identity development.

Although museum visits are highly valued by schools and teachers, it has been found that teachers spent most of their time during the visit on handling logistics, looking after the health and safety of students, or disciplining students (Tal et al. 2005; see also, Anderson et al. 2006; DeWitt and Storksdieck 2008). This suggests they may not spend sufficient time facilitating the learning process during the visit and that they may have to ensure they do so before and after the visit. Karnezou et al. (2013) noted that teachers who planned and defined learning outcomes for their students facilitated better learning, as compared to teachers who did not define learning outcomes for the visit.

Some reported developments in pre-service and in-service teacher education programs have been to educate teachers on the pedagogy required to conduct lessons at informal learning spaces (Tal et al. 2005; Sgouros and Stavrou 2019) and to help them understand their own role as promoters of students' engagement during the school visit by providing support and scaffolding to connect students' existing conceptions with the exhibits (Faria and Chagas 2013).

Teachers ought to be given the time and autonomy to plan, prepare, and execute the field trip as a teacher acts as a link between SCM visits and school science content. Additionally, more collaboration between the science museums and teachers should be encouraged, since one main reason teachers are unable to link the schools' science curriculum is that they are not fully aware of the content of the exhibits. Teachers can and must bridge the connections between school curriculum, museum content, and students' inquiry (Karnezou et al. 2013; Faria and Chagas 2013) thus ensuring the field trip is rewarding for the participants.

13.3.1.3 Families

Learning among family groups in SCMs has not been well-studied, in contrast to research on adult visitors as individuals (Davidsson 2015). However, it is known that families tend to consider interactive science centers as places of recreation as well as places of learning for their children. Science centers provide an opportunity for families to engage with science through hands-on activities, interactive exhibits, and direct experiences with scientific phenomena. It has been shown that when families visit a science center together, they cherish the social aspect of the experience and at the same time value the entertaining qualities and the learning opportunities afforded by the venue (Borun 2002).

Family engagement and interaction in SCMs are generally measured by the quality of conversations between parent and child. A study by Falk and Needham (2011) found that parents felt SCMs were able to create opportunities for them to talk to their children about science or technology, and gave their children a different learning experience. Other research (Ash 2004; Zimmerman et al. 2010; Szechter and Carey 2009; McClain and Zimmerman 2014) found that families mixed everyday conversation and scientific talk when discussing life science at SCMs. Prior experiences were leveraged by families to establish a set of mutual vocabulary between the parents and children, to ensure everyone was participating in the science discussion (Zimmerman et al. 2010). However, Gutwill and Allen (2010) reported that current interactive exhibits often do not provide enough manipulation options to promote prolonged and personalized engagement so as to keep children and parents interested in exploring and talking about a phenomenon. Tscholl and Lindgren (2016) found that the use of technology such as augmented and mixed reality technologies can offer ways to structure children's activity and to design resources to support parents in guiding their child's learning in SCMs.

The social aspect of the science experience of families in SCMs might be also related to the parents' education levels. Studies have found correlations between the education levels of parents and the amount of explanatory conversations given to their children (Szechter and Carey 2009; see also, Siegel et al. 2007). Parents who attained higher levels of education were found to spend more time with their children while interacting with the exhibits (Szechter and Carey 2009), but these parents also tend to provide explanations in a more directive manner rather than engage in interactional and collaborative discussion with their children (Siegel et al. 2007).

The socioeconomic status of families could play a role in the perception of SCMs. The way SCMs implement informal science learning can potentially socially exclude families from disadvantaged backgrounds (Dawson 2014). The assumptions of visitors' level of knowledge in science and language, could potentially drive disadvantaged families away from SCMs. One key reason could be that SCMs learning resources tend to cater to visitors from affluent, middle-class backgrounds who belong to the ethnically dominant group within the population (Bell et al. 2016).

Studies investigating language use in SCMs and minority ethnic visitors found that language mismatches between SCMs and minority ethnic visitors affected these visitors in terms of access to information and their ability to learn, or indeed be comfortable in SCMs (Ash 2004; see also, Rahm and Ash 2008; Yalowitz et al. 2015).

To conclude, parents can choose to positively influence a child. Parents who support the learning of science are crucial as it has been shown that even if families lack physical resources (e.g., low socioeconomic status families), it is still possible for their children to have high motivation to learn science and attain satisfactory achievement, as long as the children explicitly perceive support of learning from their family members (Sha et al. 2015). SCMs thus need to be made more accessible to minority and disadvantaged groups of visitors. Dawson (2014) suggests that SCMs could start by paying more attention to their current practices and make inclusiveness a key issue in science education and start designing education practices that considers the background of all visitors.

13.3.1.4 Adult Visitors

Recent studies on promoting science literacy in adult visitors focused on elucidating factors contributing to the science identity of adult visitors. Various studies have shown that science-related museums promote public understanding of science and are correlated to interest in science (Falk and Needham 2011; Falk et al. 2018; see also, Falk and Needham 2013). However, quantifying educational experience in the context of adult science learning is difficult because of various interacting factors, and thus, difficult to attribute any single variable affecting science education in adults (Falk et al. 2016a, b). Furthermore, adult visitors to SCMs come from different educational backgrounds, making it difficult to establish a baseline relating to visitors' science knowledge, opinions, and motivations (Falk and Dierking 2000; Falk and Dierking 2014).

It has become increasingly important to investigate how adults develop interest in science, as there is an indication that science interests can not only be developed during their time in school, but can be inculcated over an entire lifetime (Falk and Dierking 2010; Falk and Needham 2013). To fill this gap, Shein et al. (2019) proposed "science identity" as a construct to model the self-perception of individuals, specifically whether they considered themselves a "science person". This model is mainly to better understand how a person's science identity relates to science engagement and participation, which correlates to science learning. Their results showed correlation between higher levels of science identity with positive interest in science.

In a similar vein, to understand visitors' motivation for science, it would be important to have a common set of metrics to quantify visitor profiles. Lewalter et al. (2015) proposed latent class analysis as a method to classify visitors based on their perceived science knowledge, cognitive, and motivational variables. This measure would allow for an evidence-based way of addressing the needs in different subgroups of visitors. Regardless of motivation for visiting SCMs, visitors' interest in science were influenced by their SCM experience, and all visitors to SCMs showed higher levels of science understanding compared to non-visitors, regardless of their education and income levels (Falk et al. 2016a, b). However, given that science understanding is affected by education and income levels (as shown by Falk et al. 2016a, b; Shein et al. 2019), the focus should be placed on making SCMs more inclusive to people from various backgrounds and improve science identity among visitors.

13.3.2 Museum Educators and Staff

SCM staff whose job function focuses on education are typically known as museum educators (MEs) and they play critical roles within SCMs. They research, design, and even build the exhibits, as well as interact with a myriad of visitors of different age groups, with different backgrounds and prior experiences, possessing different learning skills and interests, and yet must be ready to engage with and educate them at varying levels. However, there is limited research in this aspect of science museums. It is only since the 2010s that there has been an increase in the literature relating to science MEs, which mainly examined museum educators' perceptions on science content and educators' interaction with visitors (Piqueras and Achiam 2019).

13.3.2.1 Interactions with Visitors

Their interactions with visitors are short, and Shaby et al. (2019b) noted that "interactions consisting of scientific explanations were very rare", which echoed findings by Tran and King (2011). MEs tend to be seen as a 'technician' who provides technical support or simply demonstrates the operations of unfamiliar exhibits (Allen 2004; Faria and Chagas 2013; Shaby et al. 2019b). But beyond explanations of scientific content or how an exhibit functions, MEs are typically also expected to include elements in their presentation or interactions with visitors that create a positive memory of the visit, in order to develop long-term interest in science (Cox-Petersen et al. 2003; Tal et al. 2005). Although the aims of science teaching are similar in schools and museums, Tran (2006) highlighted some major differences, such as the limited contact time and the MEs' lack of familiarity with the individual learner/visitor, which are obstacles to nurturing students' interest in science at SCMs. Furthermore, science teaching in SCMs is typically a one-off event, with little or no follow-up activities back in schools, making it even more difficult to sustain interest in science learning at SCMs (Anderson and Zhang 2003; Storksdieck 2001).

For visiting school groups, MEs may be able to achieve better and more consistent outcomes by cooperating with the school teachers in the group. As previously mentioned, collaboration between MEs and school teachers should be encouraged. While MEs focus on the content presented in the exhibits and help teachers prepare by providing supporting material about the exhibits, teachers can focus on bridging the exhibit content to the school curriculum and structuring the learning during the visit such that it is relevant to the learning outcomes for students (Tran 2006).

It has been suggested that to further develop MEs and their contribution to informal science learning, collaborations with universities should be encouraged for the purposes of creating professional development programs for MEs. Furthermore, such collaborations could for example, expand the research on the learning of science in informal spaces, so as to support and develop the educational agenda that is unique to informal learning in SCMs (Kelly and Kassing 2013; Kisiel et al. 2012; Penuel 2017).

13.3.2.2 Choice of Content

Clearly, choices by museum educators on the content and perspectives to include in an exhibit have an impact on how the concepts, as well as the overall image or nature of science, are communicated. Fundamentally, such choices begin with broad parameters such as the specific subject area and thematic organization of the museum but quickly requires focus on the types of specific exhibits to include insofar as issues such as whether the exhibit is pedagogical or experiential (i.e., exhibits where information is presented and passively viewed versus those where the phenomena is actively experienced), dogmatic or dialectic (i.e., where information is portrayed in a factual manner versus approaches that involve the visitor deriving understanding within their personal contexts). Research in this area serves not only to understand these processes, but can also inform practice. For example, Johnson et al. (2019) suggested that in addition to typical pedagogical approaches in exhibit design that seek to deliver and reinforce factual content, it is also important to apply other methods, such as refutation or promoting of inquiry learning, to better facilitate conceptual change. Three main areas of interest relating to choice of content could be discerned and are discussed below.

Types of Exhibits. Two papers in this review helpfully proposed classification schemes for the types of exhibits commonly found in SCMs—useful for museum staff design work as well as in exhibit-related research. Stocklmayer and Gilbert (2002) examined informal learning in the context of interactive exhibitions and framed these types of exhibits as exemplars of real-world phenomena or analogical representations of a consensus model of a phenomenon. Additionally, these exhibits could be featured together with a theme, or could be a standalone type. Hence, four general categories of interactive exhibits were identified. These were: (1) *exhibits as exemplars of real-world phenomena*, designed to provide experience of the phenomena; (2) *exhibits as showing only similarities between entities*, where exhibits that were similar to a phenomena in appearance only; (3) *exhibits showing similarities between both entities and relationships*, which generally model the phenomena and the causal mechanisms of the phenomena with an analogy; and, (4) *exhibits only showing similarities between relationships*, which encompasses exhibits that modelled causal mechanisms of a phenomena with an analogy based on the causal mechanisms of a different phenomenon.

Pedretti (2004) adopted a different approach and classified exhibits into three non-mutually exclusive categories. Unlike the approach in Stocklmayer and Gilbert (2002) where the categorizations focus on grouping by the representations of the phenomena being exhibited, Pedretti (2004) grouped exhibits by the intended learning experience presented to the visitor. The first two general types of exhibits are those that are commonly encountered in science exhibitions: the *pedagogical* category, where the exhibit is conveying a specific science finding in a manner that that is similar to conventional teaching; while the *experiential* category of exhibits is intended to convey interactional experiences to visitors. The third and emerging type of exhibit is the *critical* category, where exhibits are designed to present multiple perspectives of socio-scientific issues and provoke debate.

Emerging Trend of Socio-Scientific Exhibits. In recent years, there is an increasing use of socio-scientific exhibits to engage with the public. Socio-scientific exhibits include both scientific and societal aspects, through the use of such exhibits, it hopes to promote conversations on current science and technology. Pedretti (2004) noted that critical issues-based exhibits can be used to raise social awareness of science and technology in the sociocultural context. The way these exhibits personalize subject matter and evoke emotion can help stimulate learning. Pedretti (2004) concluded that although the personal context is prominently featured in critical issues-based exhibits, the sociocultural and physical context should not be ignored, but should be taken into consideration together with the personal context. Such methods would be important to engage visitors on polarizing topics like climate change (Hoffman 2020; Jones et al. 2020).

In a more recent study, Kollmann et al. (2013) examined two case studies that used socio-scientific issues to engage and teach the public about current science and technology topics. They noted that science content should be presented in a way that considers the visitors' prior knowledge, values, and experiences, so as to promote discussion. The approach these case studies took was to present science information from experts, and to encourage participants to discuss among themselves, which also promoted visitor-generated content. Kollmann et al. (2013) concluded that structuring museum exhibits and programs around current socio-scientific issues, and encouraging dialogue also increased visitors' confidence when participating in discussions about STEM issues. Similarly, Pedretti (2004) and Walker et al. (2013) showed that science exhibitions dealing with critical socio-scientific issues can challenge visitors in different ways and change behavior. These exhibitions were able to enhance learning by personalizing the subject, which evoked emotion and stimulated dialogue and debate. In contrast to the aforementioned trend, Davidsson and Jakobsson (2007) found that some museum staff hesitated to include controversial socio-scientific issues in exhibitions, choosing instead to present material in unproblematic ways.

Factors Influencing Exhibit Curation. Science exhibitions can also be influenced externally by sponsorships. For example, sponsors in many cases, set criteria for exhibitions at science museums, thus influencing the work of developing new exhibitions, while museum staff would also practice self-censorship to take into account what is believed to be the intention of sponsors (Davidsson and Sørensen 2010).

In an attempt to elucidate the considerations of exhibition designers, Davidsson and Jakobsson (2009) examined the process by which museum staff determine what science content to include in an exhibition. They found that museum staff typically rely on their personal experiences, coupled with professional education, experiences, and external references (literature and other communications from the broader science community), as well as some degree of pedagogical knowledge, formally and/or informally acquired as the case may be for individual staff members.

Another study by Davidsson (2009) looked into what aspects of science do museum staff consider when planning for a new exhibition. Interviews revealed that when museum staff were asked to consider presenting science in the context of society, politics, and the economy, they typically felt implicit, rather than explicit, examples of how these socio-scientific issues are represented in their exhibitions sufficed. That is, visitors would be able to notice and grasp the underlying socio-scientific relationships and perspectives, merely from the context in which the scientific content is situated. This could potentially result in controversial, or simply socio-cultural, aspects of science to be either consciously or unconsciously excluded. A study by Rennie and Williams (2002) also supported this view and they proposed that improvement should be made to explicitly represent controversial aspects of science, so that visitors can be encouraged to think critically about science.

13.3.2.3 Partnerships

In the 2010s, a slew of studies looked into the benefits of SCM-university partnership programs. Collaborative programs with universities and scientists allow museum educators to be kept up to date with the newest findings and when translated into exhibits, also gives the public an insight into significant trends in research. Silverman and Bartley (2013), Bell et al. (2016), and Trahan et al. (2019) showed such collaborations could bring scientific research to the public in compelling ways. Silverman and Bartley (2013) proposed that shared goals between the SCM and university, together with the openness to learn from each other were important factors that contributed to a successful program for visitors with disabilities. Trahan et al. (2019) also supported this mutual learning idea, by showing that it is an iterative process of design and testing of educational tools together with the SCM, to design appropriate language support in programs to support linguistically diverse students and backgrounds and improve social inclusion. Bell et al. (2016) demonstrated how the public can experience what it is like to be a researcher with joint-appointment research staff and embedded laboratories in the SCM.

Other collaborative programs looked into the promotion of science to students and training programs for future scientists. Durksen et al. (2017) emphasizes the importance of partnership between researchers and museums after a successful collaboration in a program for children. Wadman et al. (2009) described a collaborative project to help train college interns to design a material science program for middle school students. Such a partnership was able to achieve multiple goals of improving material science lessons for a middle school audience and helped college interns develop their informal education skills professionally. Itzek-Greulich and Vollmer (2017) noted that students in laboratory programs in an out-of-school environment are more motivated and have better situational competence (i.e., competency in science laboratory practical work).

13.3.3 Exhibits and Exhibit Design

Exhibits are the key communication tool between visitors and the scientific content and/or perspective being portrayed. Exhibits have to be designed effectively to bring forth science concepts, the broader nature of science, and socio-scientific issues to a broad gamut of visitor profiles but designing exhibits for accurate content while being engaging to visitors is difficult (Allen 2004). In addition, however well an exhibit was designed, it remains a possibility that a visitor may interpret the information presented in unexpected ways (Falcão et al. 2004). Of particular interest to researchers in SCMs is the meaning visitors obtain from their interactions with exhibits and how they make sense of these experiences. By understanding how visitors interpret and interact with the exhibits, exhibits be designed to improve visitors' experiences (DeWitt and Osborne 2010). Researchers analyzed the characteristics and components of exhibits which may have an impact on visitors' learning. Aside from the obvious interest in visitor engagement and their interactions, other notable themes in the literature include the use of real versus replicated exhibits, the design of labels, and the increasing use of technology in the exhibits.

13.3.3.1 Engagement and Interaction

From the papers reviewed, there is a noticeable portion of papers dedicated to engagement and interaction, including those examining visitors' social interactions. A number of these studies were dedicated to family interactions in SCMs, specifically, how family members interact to make sense of exhibits and how to facilitate such interactions (Falk and Storksdieck 2010; Gutwill and Allen 2010; McClain and Zimmerman 2014; Thomas and Anderson 2013; Zimmerman et al. 2008, 2010). Some studies examined how disadvantaged families interact in SCMs and recommended ways to support these families (Archer et al. 2016; Feinstein and Meshoulam 2014; Weiland 2015). The remaining studies on engagement and interaction attempted to understand the various factors influencing interactions (Anderson et al. 2003; see also, Rahm 2004), with one study about the balance between entertainment and scientific content (Shea 2014).

Beyond visitor engagement, there were also the questions of whether learning of relatively complex topics or imparting of scientific perspectives could be achieved in SCM settings. In general, various studies showed that this was indeed possible. In Pitts et al. (2014), they explored how complex physics can be presented to young students. It was shown that students could measurably learn about abstract concepts in gravity without losing interest or feeling intimidated by the topic when engaged in age-appropriate ways. DeWitt and Osborne (2010) used photographs and videos of how school children interact with the exhibits to trigger recollections of the concept the exhibit was conveying. These stimulated-recall interviews elicited affective and cognitive reflections of the prior visit and students could use their science understandings to offer causal explanations of what they had observed. A similar study by Medved and Oatley (2000) measured knowledge retention by examining episodic and semantic memory change after a period. They found that while episodic memories of the experienced events of the visit remained stable for nearly all visitors, semantic memories of the conceptual knowledge gained from the visit were either enhanced, remained stable, or deteriorated in approximately equal proportions of visitors. Botelho and Morais (2006) investigated how students interacted with exhibits and suggested that the learning outcome of the exhibit should be explicitly mentioned, and that the exhibit design and mechanism should fit this learning outcome in order to facilitate the learning of the desired scientific concepts.

In addition to conveying knowledge, SCM programs can also be used to understand social and cultural issues in STEM. One unique study on engagement and interaction looked into combining knowledge from indigenous people with Western scientific knowledge to promote STEM interest (Augare et al. 2017). Caleon and Subramaniam (2005) discussed the impact of a cryogenics program for middle school students, reporting changes in attitudes towards science. Pattison et al. (2018) examined how identity negotiations occur within an informal engineering setting. Findings indicated that leadership-oriented youths in the groups played an important role in influencing identity negotiation, indicating that more attention should be paid to peer learning dynamics, as it can facilitate or diminish learning.

13.3.3.2 Real vs Replica

The relation of authenticity of objects in exhibits to learning have been examined over the years, but conflicting evidence in the literature makes it difficult to arrive at any clear conclusions (Carsten Conner and Perin 2020). A study by Hampp and Schwan (2015) on the impact of authenticity of scientific objects in science museums, discussed four dimensions of authenticity. Their results suggested that the scientific or technological principle is more important than the authenticity of the object, and that the only time authenticity matters is when the historical significance of the object is brought to the foreground. Carsten Conner and Perin (2020) extended this investigation to include comparisons between objects in real and replicated environments. They found differences in the type of discussion between families in both types of environments, with more perceptual discussions (from descriptions of features or sensory-based observations) present in the real environment, as compared to the replicated environment. However, more conceptual conversations (from information communicated by the tour or exhibit itself) were recorded in the replicated environment. They suggest that visitors are better engaged by rich sensory experiences from real-world settings.

13.3.3.3 Exhibit Labels

Labels are typically used in museums to describe an exhibit or provide supporting information. Hohenstein and Tran (2007) noted the lack of studies on the influence of such labels on visitors' learning behaviors. Their findings indicated that adding a guiding question ("Why is this here?") to the label of an exhibit promoted greater discussion among visitors and could stimulate inquiry and reflection. Similarly, Phelan et al. (2017) found that adding risk–benefit information (e.g., presenting two different viewpoints on the topic) on exhibit labels could change visitors' attitudes and promoted reflection on both sides of a scientific argument.

Another study by Atkins et al. (2009) looked at how labels help visitors make sense of how an exhibit is intended to function, thereby learning about the scientific principles behind the exhibit. Their results indicated that labels were deeply influential on visitors' interactions with the exhibit and with each other, with both intended and sometimes unintended effects. The intended effect of labels is to guide visitors in their interaction with the exhibit and/or provide conceptual information for learning, and detailed labels often achieve this. On the other hand, brief, incomplete, or missing labels may unexpectedly result in visitors inventing their own interactions with the exhibits and engaging in more collaborative and meaningful scientific discussions within a group of visitors. This was echoed in the study by McClain and Zimmerman (2014) where nature trails with few educational signages saw visitors actively engaging in connecting their personal everyday experiences to what they saw during the visit, whereas visitors to trails with detailed signages mainly read and accepted what was presented. Thus, exhibits labels have to be carefully designed with the goals of the exhibits in mind, as it may not always be better to have more information or guidance.

13.3.3.4 Use of Technology

There has been an increasing use of technology in SCMs and researchers have been looking at various aspects of incorporating technology into exhibits. For example, Yoon et al. (2013) investigated the cognitive effect of different types of scaffolding on informal learning. They explored the use of augmented reality (AR) scaffolding in combination with different types of pedagogy-based scaffolding. They noted that the use of digital augmentation could have the potential to enhance conceptual understanding among student visitors. Price et al. (2015) explored the effects of new technologies on informal learning. They were interested to find out whether stereoscopic films could help in retaining learning gains. They recruited two groups of participants, where one was shown a 3D presentation, while the other acted as control and shown a 2D presentation. They found no significant difference between the two groups of participants in their recall of knowledge immediately after the presentation. However, participants in the 3D presentation group were able to retain more knowledge later, as compared to those in the 2D presentation group, suggesting that 3D presentations have a positive effect on knowledge retention.

In Salmi et al. (2017b), AR-enhanced exhibits on the Doppler effect and on the movement of molecules in gases were used in a study to evaluate how well abstract science concepts can be conveyed to students. One example of the AR sessions overlaid simulations of molecular motion with graphs of temperature and velocity. During the pretest, there was a gender effect in science knowledge, with males having higher scores than females. However, in the posttest after the AR session, the gender effect disappeared, which suggested that girls may gain more than boys when using AR. AR was also shown to be the most beneficial for the lowest achieving students, considering that all students achieved better scores in the knowledge test after the AR session. The use of AR thus appears to have a levelling-up effect on learning from such exhibits.

13.4 Conclusion

The aim of this chapter was to summarize research conducted between 2000 and 2020 on science education in SCMs. It is therefore a review of educational research in the area of informal science learning associated with science centers and museums, on visitors' perceptions and experiences, the effectiveness of exhibits in terms of engagement and interaction, the factors that influence the curation and design of exhibits, and other related aspects of the literature within this milieu. However, it should be noted that the majority of these high-quality studies were situated in highly industrialized nations and thus may not always be representative of SCMs from other cultural and economic circumstances. This represents an opportunity for such studies to be conducted in these other contexts.

Overall, it was clear that the informal learning environment of SCMs has enabled and fostered visitors' interest in and development of various dimensions of scientific knowledge, and SCMs have continued to progressively refine their approaches towards this goal. In many ways, the reviewed research represented a microcosm of the three historical waves of learning research trends described by Crowley et al. (2014) through the successive use of behaviorist, constructivist, and sociocultural lenses to examine and understand how and what visitors learn in SCMs. Though not distinctly in "waves", we could discern work that attempted to empirically determine the characteristics of exhibits that tended to attract visitor attention, as well as those papers that centered on the use of theoretical perspectives to examine various aspects of a visit to SCMs. There was also a general evolution towards an increased adoption of exhibits that aimed to present relevant socio-scientific issues alongside traditional scientific knowledge to raise awareness and encourage conversations among visitors, although there still appears to be a reticence amongst museum educators to make such links too explicit, perhaps for fear of provoking too much controversy. Another area in which SCMs have been identified as working towards but still with much room to develop is that of inclusivity. In other words, to welcome and cater to visitors regardless of background, especially those disadvantaged in real or self-perceived ways in terms of socioeconomic circumstances, educational levels, or readiness to engage and interact with the exhibits. Understandings derived from the research, of how and why museum educators conceive, create, and curate their exhibits can help ameliorate these issues by raising awareness of blind spots, even as they help highlight what works. Indeed, students, teachers, family groups, and adult visitors all hold generally positive recollections and attitudes towards SCMs.

As public spaces that promote the learning of science in its various forms, and increasingly in raising awareness of important socio-scientific issues, SCMs not only serve to indirectly complement and directly support formal science learning in schools through SCM-teacher cooperation, they fundamentally broaden access and equity in science learning too. It is not yet clear though, if SCMs and their affordances for informal science learning will eventually translate into tangible uptake of science and STEM as subject specializations and careers.

13.5 Note to Our Future Colleagues

Dear future colleagues,

Our review sought to encapsulate the state of research in this field in the first 20 years of this century, where it might perhaps be of some use in guiding further research between the time of this writing and the time when you are reading this. It would also be by then, a somewhat distant reference to look back upon and compare notes with your present. We sincerely hope such a comparison will make you gasp in wonder at the progress made, barriers long since dismantled, and perhaps what might be to us unimaginable paradigm shifts that have advanced the field and brought about huge leaps in the learning of science in informal contexts.

From our perspective, looking ahead towards 2050, we expect science centers to become even more immersive and interactive, leveraging emerging technologies like virtual and augmented reality, artificial intelligence, and robotics to engage visitors in new and exciting ways. Exhibits may be more personalized and adaptive, tailored learning experiences designed to cater to individual interests and learning preferences. In this imagined and desirable future, science centers have become more integrated with the community, functioning not just as places for learning, but also as open communal hubs for integrative STEM collaboration and innovation. That they may partner with local businesses, organizations, and even interested individuals to develop new technologies and solutions to real-world problems, while also making these developmental processes and activities exhibits in and of themselves. Such processes and activities have hitherto remained out of the public eye. Putting them on display would reveal the applicability, relevance, and impact of science, technology, engineering, and mathematics to the lay person.

Another important trend we see in the science center of the future is a greater emphasis on sustainability and environmental stewardship. As the world faces growing environmental challenges, science centers must seize opportunities to educate the public on the importance of sustainable practices and inspire them to take action.

Yet, despite all of these changes and advancements, the core mission of science centers will surely remain the same: to promote scientific literacy, inspire curiosity, and provide accessible and engaging learning experiences for all. This pivotal role will certainly continue to grow in importance as our already VUCA (volatile, uncertain, complex, and ambiguous) world becomes ever more so. It is therefore imperative that we remind you, as we remind ourselves, to keep our collective finger on the pulse of society in terms of its development and hence its concomitant appetite for knowledge, the emerging trends in technology, not only for its potential but also the challenges and risks that new and disruptive innovations may bring, and perhaps most crucially, how, when, and where we can situate and insinuate the learning of science in our everyday lives. As we look ahead to the future, we must ensure that these fundamental principles remain at the heart of everything we do, and we hope you will continue to build on our efforts.

References

Anderson D, Kisiel J, Storksdieck M (2006) Understanding teachers' perspectives on field trips: discovering common ground in three countries. Curator Museum J 49(3):365–386. https://doi.org/10.1111/j.2151-6952.2006.tb00229.x

Anderson D, Zhang Z (2003) Teacher perceptions of field-trip planning and implementation. Visitor Stud Today! 6(3):6–11

Archer L, DeWitt J, Willis B (2014) Adolescent boys' science aspirations: Masculinity, capital, and power. J Res Sci Teach 51(1):1–30. https://doi.org/10.1002/tea.21122

Augare HJ, Davíd-Chavez DM, Groenke FI, Plume-Weatherwax ML, Fight LL, Meier G et al (2017) A cross-case analysis of three native science field centers. Cult Stud Sci Educ 12(2). https://doi.org/10.1007/s11422-015-9720-6

Borun M (2002) Object-based learning and family groups. In: Paris SG (ed) Perspectives on object-centered learning in museums. L. Erlbaum Associates, Mahwah, N.J., pp 245–261

Cox-Petersen AM, Marsh DD, Kisiel J, Melber LM (2003) Investigation of guided school tours, student learning, and science reform recommendations at a museum of natura history. J Res Sci Teach 40(2):200–218. https://doi.org/10.1002/tea.10072

Crowley K, Pierroux P, Knutson K (2014) Informal learning in museums. In: Sawyer RK (ed), 2nd edn. Cambridge University Press, pp 461–478. https://doi.org/10.1017/CBO9781139519526.028

Davidsson E (2015) Investigating visitors' learning related to science centre exhibits—a progress report of recent research literature and possible future research foci. Utbildning & Larande 6(2):28–47

Davidsson E, Sørensen H (2010) Sponsorship and exhibitions at Nordic science centres and museums. Museum Manage Curatorship 25(4):345–360. https://doi.org/10.1080/09647775.2010.525399

DeWitt J, Storksdieck M (2008) A Short review of school field trips: key findings from the past and implications for the future. Visitor Stud 11(2):181–197. https://doi.org/10.1080/10645570802355562

Eshach H (2007) Bridging in-school and out-of-school learning: formal, non-formal, and informal education. J Sci Educ Technol 16(2):171–190. https://doi.org/10.1007/s10956-006-9027-1

Falk JH, Dierking LD (2000) Learning from museums: visitor experiences and the making of meaning. AltaMira Press

Falk J, Dierking L (2010) The 95 percent solution: school is not where most Americans learn most of their science. Am Sci 98:486–493. https://doi.org/10.1511/2010.87.486

Falk JH, Dierking LD (2014) Museum experience revisited. Left Coast Press

Falk JH, Needham MD (2013) Factors contributing to adult knowledge of science and technology. J Res Sci Teach 50(4):431–452. https://doi.org/10.1002/tea.21080

Falk JH, Staus N, Dierking LD, Penuel W, Wyld J, Bailey D (2016a) Understanding youth STEM interest pathways within a single community: the Synergies project. Int J Sci Educ Part B 6(4):369–384. https://doi.org/10.1080/21548455.2015.1093670

Kelly LA, Kassing S (2013) Complementary expertise in a zoo educator professional development event contributes to the construction of understandings of affective transformation. Cult Sci Edu 8:571–586. https://doi.org/10.1007/s11422-013-9496-5

Lai ER, Viering MM (2012) Assessing 21st century skills: integrating research findings national council on measurement in education

Laurillard D (2009) The pedagogical challenges to collaborative technologies. Int J Computer-Supported Collaborative Learn 4(1):5–20. https://doi.org/10.1007/s11412-008-9056-2

Moher D, Shamseer L, Clarke M, Ghersi D, Liberati A, Petticrew M, Shekelle P, Stewart LA, PRISMA-P Group (2015) Preferred reporting items for systematic review and meta-analysis protocols (PRISMA-P) 2015 statement. Syst Rev 4, 1. https://doi.org/10.1186/2046-4053-4-1

Penuel WR (2017) Research–practice partnerships as a strategy for promoting equitable science teaching and learning through leveraging everyday science. Sci Educ 101(4):520–525. https://doi.org/10.1002/sce.21285

Rahm J, Ash D (2008) Learning environments at the margin: case studies of disenfranchised youth doing science in an aquarium and an after-school program. Learn Environ Res 11(1):49–62. https://doi.org/10.1007/s10984-007-9037-9

Sha L, Schunn C, Bathgate M (2015) Measuring choice to participate in optional science learning experiences during early adolescence. J Res Sci Teach 52(5):686–709. https://doi.org/10.1002/tea.21210

Siegel DR, Esterly J, Callanan MA, Wright R, Navarro R (2007) Conversations about science across activities in Mexican-descent families. Int J Sci Educ 29(12):1447–1466. https://doi.org/10.1080/09500690701494100

Storksdieck M (2001) Differences in teachers' and students' museum field-trip experiences. Visitor Stud Today! 4(1):8–12

Tran L, King H (2011) Teaching Science in informal environments: pedagogical knowledge for informal educators, 279–293

Yalowitz SS, Garibay C, Renner N, Plaza C (2015) Bilingual Spanish-English intergenerational groups' experiences in bilingual exhibitions. Museums Soc Issues 10(1):35–51. https://doi.org/10.1179/1559689314Z.00000000031

Bibliography

Achiam MF (2013) A content-oriented model for science exhibit engineering. Int J Sci Educ Part B 3:214–232. https://doi.org/10.1080/21548455.2012.698445

Afonso AS, Gilbert JK (2006) The use of memories in understanding interactive science and technology exhibits. Int J Sci Educ 28(13):1523–1544. https://doi.org/10.1080/09500690600768828

Afonso AS, Gilbert JK (2007) Educational value of different types of exhibits in an interactive science and technology center. Sci Educ 91(6):967–987. https://doi.org/10.1002/sce.20220

Afonso AS, Gilbert JK (2008) The nature of exhibits about acoustics in science and technology centres. Res Sci Educ 38:633–651

Alexander JM, Johnson KE, Kelley K (2012) Longitudinal analysis of the relations between opportunities to learn about science and the development of interests related to science. Sci Educ 96(5):763–786. https://doi.org/10.1002/sce.21018

Allen S (2004) Designs for learning: studying science museum exhibits that do more than entertain. Sci Educ 88(S1):S17–S33. https://doi.org/10.1002/sce.20016

Anderson D, Lucas KB, Ginns IS (2003) Theoretical perspectives on learning in an informal setting. J Res Sci Teach 40(2):177–199. https://doi.org/10.1002/tea.10071

Archer L, Dawson E, Seakins A, Wong B (2016) Disorientating, fun or meaningful? Disadvantaged families' experiences of a science museum visit. Cult Stud Sci Educ 11(4):917–939. https://doi.org/10.1007/s11422-015-9667-7

Ash D (2004) Reflective scientific sense-making dialogue in two languages: the science in the dialogue and the dialogue in the science. Sci Educ 88(6):855–884. https://doi.org/10.1002/sce.20002

Atkins LJ, Velez L, Goudy D, Dunbar KN (2009) The unintended effects of interactive objects and labels in the science museum. Sci Educ 93(1):161–184. https://doi.org/10.1002/sce.20291

Bell J, Chesebrough D, Cryan J, Koster E (2016) Museum-university partnerships as a new platform for public engagement with scientific research. J Museum Educ 41(4):293–306. https://doi.org/10.1080/10598650.2016.1228302

Botelho A, Morais AM (2006) Students–exhibits interaction at a science center. J Res Sci Teach 43(10):987–1018. https://doi.org/10.1002/tea.20135

Braund M, Reiss M (2006) Towards a more authentic science curriculum: the contribution of out-of-school learning. Int J Sci Educ 28(12):1373–1388. https://doi.org/10.1080/09500690500498419

Briseño-Garzón A (2013) More than science: family learning in a Mexican science museum. Cult Sci Edu 8:307–327. https://doi.org/10.1007/s11422-012-9477-0

Brooke H, Solomon J (2001) Passive visitors or independent explorers: responses of pupils with severe learning difficulties at an interactive science centre. Int J Sci Educ 23:941–953. https://doi.org/10.1080/09500690010016094

Bunten A, Arvizu S (2013) Turning visitors into citizens: using social science for civic engagement in informal science education centers. J Museum Educ 38:260–272. https://doi.org/10.1080/10598650.2013.11510778

Burke LE-A (2020) Informal science educators and children in a low-income community describe how children relate to out-of-school science education. Int J Sci Educ 42(10):1673–1696. https://doi.org/10.1080/09500693.2020.1774936

Caleon I, Subramaniam R (2005) The impact of a cryogenics-based enrichment programme on attitude towards science and the learning of science concepts. Int J Sci Educ 27:679–704. https://doi.org/10.1080/09500690500038306

Carsten Conner LD, Perin SM (2020) Learning from the real versus the replicated: a comparative study. Int J Sci Educ Part B 10(3):266–276. https://doi.org/10.1080/21548455.2020.1831707

Dairianathan A, Subramaniam R (2011) Learning about inheritance in an out-of-school setting. Int J Sci Educ 33:1079–1108. https://doi.org/10.1080/09500693.2010.494185

Davidsson E (2009) Enhancing visitors' interest in science—a possibility or a paradox? Res Sci Educ 39(2):197–213. https://doi.org/10.1007/s11165-008-9080-2

Davidsson E, Jakobsson A (2007) Different images of science at Nordic science centres. Int J Sci Educ 29(10):1229–1244. https://doi.org/10.1080/09500690600969848

Davidsson E, Jakobsson A (2009) Staff members' ideas about visitors' learning at science and technology centres. Int J Sci Educ 31(1):129–146. https://doi.org/10.1080/09500690701649588

Dawson E (2014) "Not designed for us": how science museums and science centers socially exclude low-income, minority ethnic groups. Sci Educ 98(6):981–1008. https://doi.org/10.1002/sce.21133

DeWitt J, Osborne J (2007) Supporting Teachers on science-focused school trips: towards an integrated framework of theory and practice. Int J Sci Educ 29(6):685–710. https://doi.org/10.1080/09500690600802254

DeWitt J, Osborne J (2010) Recollections of exhibits: stimulated-recall interviews with primary school children about science centre visits. Int J Sci Educ 32(10):1365–1388. https://doi.org/10.1080/09500690903085664

Durksen TL, Martin AJ, Burns EC, Ginns P, Williamson D, Kiss J (2017) Conducting research in a medical science museum: lessons learned from collaboration between researchers and museum educators. J Museum Educ 42(3):273–283. https://doi.org/10.1080/10598650.2017.1339171

Essex J, Haxton K (2018) Characterising patterns of engagement of different participants in a public STEM-based analysis project. Int J Sci Educ Part B 8:178–191. https://doi.org/10.1080/21548455.2017.1423128

Falcão D, Colinvaux D, Krapas S, Querioz G, Alves F, Cazelli S et al (2004) A model-based approach to science exhibition evaluation: a case study in a Brazilian astronomy museum. Int J Sci Educ 26(8):951–978. https://doi.org/10.1080/1468181032000158363

Falk JH, Needham MD (2011) Measuring the impact of a science center on its community. J Res Sci Teach 48(1):1–12. https://doi.org/10.1002/tea.20394

Falk J, Storksdieck M (2005) Using the contextual model of learning to understand visitor learning from a science center exhibition. Sci Educ 89(5):744–778. https://doi.org/10.1002/sce.20078

Falk JH, Storksdieck M (2010) Science learning in a leisure setting. J Res Sci Teach 47(2):194–212. https://doi.org/10.1002/tea.20319

Falk JH, Dierking LD, Swanger LP, Staus N, Back M, Barriault C et al (2016b) Correlating science center use with adult science literacy: an international cross-institutional study. Sci Educ 100(5):849–876. https://doi.org/10.1002/sce.21225

Falk JH, Pattison S, Meier D, Bibas D, Livingston K (2018) The contribution of science-rich resources to public science interest. J Res Sci Teach 55(3):422–445. https://doi.org/10.1002/tea.21425

Faria C, Chagas I (2013) Investigating school-guided visits to an aquarium: what roles for science teachers? Int J Sci Educ Part B 3(2):159–174. https://doi.org/10.1080/09500693.2012.674652

Feinstein NW, Meshoulam D (2014) Science for what public? Addressing equity in American science museums and science centers. J Res Sci TeAch 51(3):368–394. https://doi.org/10.1002/tea.21130

Griffin J (2004) Research on students and museums: looking more closely at the students in school groups. Sci Educ 88(S1):S59–S70. https://doi.org/10.1002/sce.20018

Guisasola J, Solbes J, Barragues J-I, Morentin M, Moreno A (2009) Students' understanding of the special theory of relativity and design for a guided visit to a science museum. Int J Sci Educ 31:2085–2104. https://doi.org/10.1080/09500690802353536

Gupta R, Voiklis J, Rank SJ, Dwyer J, Fraser J, Flinner K, Nock K (2020) Public perceptions of the STEM learning ecology—perspectives from a national sample in the US. Int J Sci Educ Part B 10:112–126. https://doi.org/10.1080/21548455.2020.1719291

Gutwill JP (2008) Challenging a common assumption of hands-on exhibits. J Museum Educ 33:187–198. https://doi.org/10.1080/10598650.2008.11510599

Gutwill JP, Allen S (2010) Facilitating family group inquiry at science museum exhibits. Sci Educ 94(4):710–742. https://doi.org/10.1002/sce.20387

Hampp C, Schwan S (2015) The role of authentic objects in museums of the history of science and technology: findings from a visitor study. Int J Sci Educ Part B 5(2):161–181. https://doi.org/10.1080/21548455.2013.875238

Heard PF, Divall SA, Johnson SD (2000) Can 'ears-on' help hands-on science learning for girls and boys? Int J Sci Educ 22:1133–1146. https://doi.org/10.1080/09500690050166715

Henriksen EK, Jensen F, Sjaastad J (2015) The role of out-of-school experiences and targeted recruitment efforts in Norwegian science and technology students' educational choice. Int J Sci Educ Part B 5:203–222. https://doi.org/10.1080/21548455.2014.900585

Hoffman JS (2020) Learn, prepare, act: "throwing shade" on climate change. J Museum Educ 45:28–41. https://doi.org/10.1080/10598650.2020.1711496

Hohenstein J, Tran LU (2007) Use of questions in exhibit labels to generate explanatory conversation among science museum visitors. Int J Sci Educ 29(12):1557–1580. https://doi.org/10.1080/09500690701494068

Itzek-Greulich H, Vollmer C (2017) Emotional and motivational outcomes of lab work in the secondary intermediate track: the contribution of a science center outreach lab. J Res Sci Teach 54(1):3–28. https://doi.org/10.1002/tea.21334

Johnson ML, Chekour A, Vaughn AR, Taasoobshirazi G (2019) Softening the landing: approaches to facilitating conceptual change for science museum educators. J Museum Educ 44:325–331. https://doi.org/10.1080/10598650.2019.1621647

Jones R, Hussain N, Spiewak M (2020) The Critical role research and evaluation assume in the post-truth era of climate change. J Museum Educ 45:64–73. https://doi.org/10.1080/10598650.2020.1720402

Kang C, Anderson D, Wu X (2010) Chinese perceptions of the interface between school and museum education. Cult Sci Edu 5:665–684. https://doi.org/10.1007/s11422-009-9197-2

Karnezou M, Avgitidou S, Kariotoglou P (2013) Links between teachers' beliefs and their practices in a science and technology museum visit. Int J Sci Educ Part B 3(3):246–266. https://doi.org/10.1080/21548455.2013.773467

Kazama T, Ogawa M (2015) Content analysis of life exhibitions in Japanese science museums and centres. Int J Sci Educ Part B 5:223–249. https://doi.org/10.1080/21548455.2014.928757

Kim M, Dillon J, Song J (2020) The factors and features of museum fatigue in science centres felt by Korean students. Res Sci Educ 50(2):419–436. https://doi.org/10.1007/s11165-018-9695-x

Kisiel J, Rowe S, Vartabedian MA, Kopczak C (2012) Evidence for family engagement in scientific reasoning at interactive animal exhibits. Sci Educ 96(6):1047–1070. https://doi.org/10.1002/sce.21036

Kollmann EK, Reich C, Bell L, Goss J (2013) Tackling tough topics: using socio-scientific issues to help museum visitors participate in democratic dialogue and increase their understandings of current science and technology. J Museum Educ 38(2):174–186. https://doi.org/10.1080/10598650.2013.11510768

Krange I, Silseth K, Pierroux P (2020) Peers, teachers and guides: a study of three conditions for scaffolding conceptual learning in science centers. Cult Sci Edu 15:241–263. https://doi.org/10.1007/s11422-018-9905-x

Laherto A (2013) Informing the development of science exhibitions through educational research. Int J Sci Educ Part B 3(2):121–143. https://doi.org/10.1080/21548455.2012.694490

Lelliott A (2014) Understanding gravity: the role of a school visit to a science centre. Int J Sci Educ Part B 4(4):305–322. https://doi.org/10.1080/21548455.2013.818260

Lewalter D, Phelan S, Geyer C, Specht I, Grüninger R, Schnotz W (2015) Investigating visitor profiles as a valuable addition to museum research. Int J Sci Educ Part B 5(4):357–374. https://doi.org/10.1080/21548455.2015.1035688

Luehmann AL, Markowitz D (2007) Science teachers' perceived benefits of an out-of-school enrichment programme: identity needs and university affordances. Int J Sci Educ 29:1133–1161. https://doi.org/10.1080/09500690600944429

Marandino M (2016) The expositive discourse as pedagogical discourse: studying recontextualization in the production of a science museum exhibition. Cult Sci Edu 11:481–514. https://doi.org/10.1007/s11422-014-9625-9

Martin AJ, Durksen TL, Williamson D, Kiss J, Ginns P (2016) The role of a museum-based science education program in promoting content knowledge and science motivation. J Res Sci Teach 53(9):1364–1384. https://doi.org/10.1002/tea.21332

Martin LM (2004) An emerging research framework for studying informal learning and schools. Sci Educ 88(S1):S71–S82. https://doi.org/10.1002/sce.20020

Massarani L, Poenaru LM, Rocha JN, Rowe S, Falla S (2019) Adolescents learning with exhibits and explainers: the case of Maloka. Int J Sci Educ Part B 9:253–267. https://doi.org/10.1080/21548455.2019.1646439

McClain LR, Zimmerman HT (2014) Prior experiences shaping family science conversations at a nature center. Sci Educ 98(6):1009–1032. https://doi.org/10.1002/sce.21134

McKinnon M, Lamberts R (2014) Influencing science teaching self-efficacy beliefs of primary school teachers: a longitudinal case study. Int J Sci Educ Part B 4:172–194. https://doi.org/10.1080/21548455.2013.793432

Medved MI, Oatley K (2000) Memories and scientific literacy: remembering exhibits from a science centre. Int J Sci Educ 22(10):1117–1132. https://doi.org/10.1080/095006900429475

Meisner R, vom Lehn D, Heath C, Burch A, Gammon B, Reisman M (2007) Exhibiting performance: co-participation in science centres and museums. Int J Sci Educ 29:1531–1555. https://doi.org/10.1080/09500690701494050

Mortensen MF (2011) Analysis of the educational potential of a science museum learning environment: visitors' experience with and understanding of an immersion exhibit. Int J Sci Educ 33:517–545. https://doi.org/10.1080/09500691003754589

Murmann M, Avraamidou L (2014) Animals, emperors, senses: exploring a story-based learning design in a museum setting. Int J Sci Educ Part B 4:66–91. https://doi.org/10.1080/21548455.2012.760857

Nelson AG, Cohn S (2015) Data collection methods for evaluating museum programs and exhibitions. J Museum Educ 40(1):27–36. https://doi.org/10.1080/10598650.2015.11510830

Nesmith SM, Wynveen CJ, Dixon EM, Brooks BW, Matson CW, Hockaday WC et al (2016) Exploring educators' environmental education attitudes and efficacy: insights gleaned from a Texas Wetland Academy. Int J Sci Educ Part B 6:303–324. https://doi.org/10.1080/21548455.2015.1078519

O'Connell KB, Keys B, Storksdieck M, Rosin M (2020) Context matters: using art-based science experiences to broaden participation beyond the choir. Int J Sci Educ Part B 10:166–185. https://doi.org/10.1080/21548455.2020.1727587

Ogawa RT, Loomis M, Crain R (2009) Institutional history of an interactive science center: the founding and development of the Exploratorium. Sci Educ 93:269–292. https://doi.org/10.1002/sce.20299

Pattison SA, Gontan I, Ramos-Montañez S, Moreno L (2018) Identity negotiation within peer groups during an informal engineering education program: the central role of leadership-oriented youth. Sci Educ 102(5):978–1006. https://doi.org/10.1002/sce.21459

Pedretti EG (2004) Perspectives on learning through research on critical issues-based science center exhibitions. Sci Educ 88(S1):S34–S47. https://doi.org/10.1002/sce.20019

Peleg R, Baram-Tsabari A (2016) Understanding producers' intentions and viewers' learning outcomes in a science museum theater play on evolution. Res Sci Educ 46:715–741. https://doi.org/10.1007/s11165-015-9477-7

Phelan S, Specht I, Schnotz W, Lewalter D (2017) Attitude change when presenting science museum visitors with risk–benefit information. Sci Educ 101(6):873–886. https://doi.org/10.1002/sce.21296

Piqueras J, Achiam M (2019) Science museum educators' professional growth: dynamics of changes in research–practitioner collaboration. Sci Educ 103(2):389–417. https://doi.org/10.1002/sce.21495

Pitts M, Venville G, David ZM (2014) An exploratory study to investigate the impact of an enrichment program on aspects of Einsteinian physics on year 6 students. Res Sci Educ 44(3):363–388. https://doi.org/10.1007/s11165-013-9386-6

Price CA, Lee H-S, Subbarao M, Kasal E, Aguilera J (2015) Comparing short- and long-term learning effects between stereoscopic and two-dimensional film at a planetarium. Sci Educ 99(6):1118–1142. https://doi.org/10.1002/sce.21185

Rader KA (2019) From watchful grasshoppers to rat basketball: pedagogical lessons from the history of live animal displays in science museums. J Museum Educ 44:368–378. https://doi.org/10.1080/10598650.2019.1663684

Rahm J (2004) Multiple modes of meaning-making in a science center. Sci Educ 88(2):223–247. https://doi.org/10.1002/sce.10117

Rennie LJ, Williams GF (2002) Science centers and scientific literacy: promoting a relationship with science. Sci Educ 86(5):706–726. https://doi.org/10.1002/sce.10030

Rennie LJ, Williams GF (2006) Adults' learning about science in free-choice settings. Int J Sci Educ 28:871–893. https://doi.org/10.1080/09500690500435387

Salmi H, Thuneberg H, Vainikainen M-P (2017a) Learning with dinosaurs: a study on motivation, cognitive reasoning, and making observations. Int J Sci Educ Part B 7(3):203–218. https://doi.org/10.1080/21548455.2016.1200155

Salmi H, Thuneberg H, Vainikainen M-P (2017b) Making the invisible observable by augmented reality in informal science education context. Int J Sci Educ Part B 7(3):253–268. https://doi.org/10.1080/21548455.2016.1254358

Sandifer C (2003) Technological novelty and open-endedness: two characteristics of interactive exhibits that contribute to the holding of visitor attention in a science museum. J Res Sci Teach 40:121–137. https://doi.org/10.1002/tea.10068

Şentürk E, Özdemir ÖF (2014) The effect of science centres on students' attitudes towards science. Int J Sci Educ Part B 4(1):1–24. https://doi.org/10.1080/21548455.2012.726754

Sgouros G, Stavrou D (2019) Teachers' professional development in nanoscience and nanotechnology in the context of a community of learners. Int J Sci Educ 41(15):2070–2093. https://doi.org/10.1080/09500693.2019.1659521

Shaby N, Assaraf OB-Z, Tal T (2019a) A student's-eye view: what 4th grade students describing their visit to a science museum recall as significant. Res Sci Educ 49(6):1625–1645. https://doi.org/10.1007/s11165-017-9669-4

Shaby N, Assaraf OB-Z, Tal T (2019b) An examination of the interactions between museum educators and students on a school visit to science museum. J Res Sci Teach 56:211–239. https://doi.org/10.1002/tea.21476

Shaby N, Assaraf OB-Z, Tal T (2019c) 'I know how it works!' student engagement with exhibits in a science museum. Int J Sci Educ Part B 9:233–252. https://doi.org/10.1080/21548455.2019.1624991

Shea M (2014) The hands-on model of the internet: engaging diverse groups of visitors. J Museum Educ 39(2):216–226. https://doi.org/10.1080/10598650.2014.11510812

Shein PP, Falk JH, Li Y-Y (2019) The role of science identity in science center visits and effects. Sci Educ 103(6):1478–1492. https://doi.org/10.1002/sce.21535

Silfver E (2019) Gender performance in an out-of-school science context. Cult Stud Sci Educ 14(1):139–155. https://doi.org/10.1007/s11422-017-9851-z

Silverman F, Bartley B (2013) Who is educating whom? Two-way learning in museum/university partnerships. J Museum Educ 38(2):154–163. https://doi.org/10.1080/10598650.2013.11510766

Stocklmayer S, Gilbert JK (2002) New experiences and old knowledge: towards a model for the personal awareness of science and technology. Int J Sci Educ 24(8):835–858. https://doi.org/10.1080/09500690210126775

Suter LE (2014) Visiting science museums during middle and high school: a longitudinal analysis of student performance in science. Sci Educ 98(5):815–839. https://doi.org/10.1002/sce.21116

Szechter LE, Carey EJ (2009) Gravitating toward science: parent–child interactions at a gravitational-wave observatory. Sci Educ 93(5):846–858. https://doi.org/10.1002/sce.20333

Tal R, Bamberger Y, Morag O (2005) Guided school visits to natural history museums in Israel: teachers' roles. Sci Educ 89(6):920–935. https://doi.org/10.1002/sce.20070

Thomas GP, Anderson D (2013) Parents' metacognitive knowledge: influences on parent–child interactions in a science museum setting. Res Sci Educ 43(3):1245–1265. https://doi.org/10.1007/s11165-012-9308-z

Trahan L, Romero V, Blinderman E (2019) From the classroom to the floor: applying language supports across museum contexts. J Museum Educ 44(4):418–426. https://doi.org/10.1080/10598650.2019.1620077

Tran LU (2006) Teaching science in museums: the pedagogy and goals of museum educators. Sci Educ 91(2):278–297. https://doi.org/10.1002/sce.20193

Tscholl M, Lindgren R (2016) Designing for learning conversations: how parents support children's science learning within an immersive simulation. Sci Educ 100(5):877–902. https://doi.org/10.1002/sce.21228

Van Schijndel TJ, Raijmakers ME (2016) Parent explanation and preschoolers' exploratory behavior and learning in a shadow exhibition. Sci Educ 100:153–178. https://doi.org/10.1002/sce.21193

Van Schijndel TJ, Franse RK, Raijmakers ME (2010) The exploratory behavior scale: assessing young visitors' hands-on behavior in science museums. Sci Educ 94:794–809. https://doi.org/10.1002/sce.20394

Wadman M, Driscoll WD, Kurzawa E (2009) Creating communicative scientists. J Museum Educ 34(1):41–53. https://doi.org/10.1080/10598650.2009.11510618

Walker GJ, Stocklmayer SM, Grant WJ (2013) Science theatre: changing South African students' intended behaviour towards HIV AIDS. Int J Sci Educ Part B 3(2):101–120. https://doi.org/10.1080/09500693.2011.633939

Weiland I (2015) An exploration of hispanic mothers' culturally sustaining experiences at an informal science center. J Res Sci Teach 52(1):84–106. https://doi.org/10.1002/tea.21190

Yoon SA, Elinich K, Wang J, Schooneveld JB, Anderson E (2013) Scaffolding informal learning in science museums: how much is too much? Sci Educ 97(6):848–877. https://doi.org/10.1002/sce.21079

Zimmerman HT, Reeve S, Bell P (2008) Distributed expertise in a science center. J Museum Educ 33(2):143–152. https://doi.org/10.1080/10598650.2008.11510595

Zimmerman HT, Reeve S, Bell P (2010) Family sense-making practices in science center conversations. Sci Educ 94(3):478–505. https://doi.org/10.1002/sce.20374

The manufacturer's authorised representative in the EU is Springer Nature Customer Service Centre GmbH, Europaplatz 3, 69115 Heidelberg, Germany. If you have any concerns regarding our products, please contact ProductSafety@springernature.com

Printed and bound by CPI Group (UK) Ltd, Croydon, CR0 4YY
01/10/2026
02225855-0003